Liberty Bell 7

The Suborbital Mercury Flight of Virgil I. Grissom

Other Springer-Praxis books
of related interest by Colin Burgess

NASA's Scientist-Astronauts
with David J. Shayler
2006
ISBN 978-0-387-21897-7

Animals in Space: From Research Rockets to the Space Shuttle
with Chris Dubbs
2007
ISBN 978-0-387-36053-9

The First Soviet Cosmonaut Team: Their Lives, Legacies and Historical Impact
with Rex Hall, M.B.E.
2009
ISBN 978-0-387-84823-5

Selecting the Mercury Seven: The Search for America's First Astronauts
2011
ISBN 978-1-4419-8404-3

Moon Bound: Choosing and Preparing NASA's Lunar Astronauts
2013
ISBN 978-1-4614-3854-0

Freedom 7: The Historic Flight of Alan B. Shepard, Jr.
2014
ISBN 978-3-3190-1155-4

Colin Burgess

Liberty Bell 7

The Suborbital Mercury Flight
of Virgil I. Grissom

 Springer

Colin Burgess
Bonnet Bay
New South Wales, Australia

SPRINGER-PRAXIS BOOKS IN SPACE EXPLORATION

ISBN 978-3-319-04390-6   ISBN 978-3-319-04391-3 (eBook)
DOI 10.1007/978-3-319-04391-3
Springer Cham Heidelberg New York Dordrecht London

Library of Congress Control Number: 2014932671

Cover design: Jim Wilkie
Project copy editor: David M. Harland
Typesetting: SPi Global.

Printed on acid-free paper

Springer is part of Springer Science+Business Media (www.springer.com)

Contents

This book is respectfully and fondly dedicated to the memory and many extraordinary accomplishments of naval aviator and test pilot, Mercury astronaut and aquanaut Scott Carpenter (1 May 1925 – 10 October 2013), who on 24 May 1962 became the second American to fly into Earth orbit on the MA-7 mission aboard spacecraft Aurora 7.

Godspeed, Scott Carpenter

Our civilization is no more than the sum of all the dreams that earlier ages have brought to fulfillment. And so it must always be, for if men cease to dream, if they turn their backs upon the wonder of the Universe, the story of our race will be coming to an end.

Sir Arthur C. Clarke, CBE
(16 December 1917 – 19 March 2008)

Foreword

Where to begin? There has been a tremendous amount of material generated regarding my brother Gus, from the early days of the space program through the Apollo 1 tragedy, but I will always talk to anyone who wants to talk about him. I can tell people who he was.

What I remember most of all about Gus was the thoroughness with which he approached everything he did, and this carried over into many things – even those not related to flying. But to know about Gus, it is important to also know about our parents, Dennis and Cecile. Dad worked for the Baltimore and Ohio Railroad for 47 years, as a signal maintainer. He was one of the fortunate few who had a job during the Depression. Our parents were very giving and generous people. Although they had modest means they were always very willing to share what they had with others in need. It seemed that when I was growing up there was always a relative living with us.

I was one of four children (Gus was the oldest) and we were blessed with parents who exhibited emotional stability and a sense of security. We were all born and raised in Mitchell, Indiana, and lived in the same house until we left home. That house at 715 West Grissom Avenue – it was Baker Street until it was named after Gus – is now in the process of restoration to become a museum.

We all attended Mitchell High School. Surprisingly, Gus was not an outstanding student in high school. In fact, he probably would have been classified as an underachiever. The high school principal did not endorse his application to enter Purdue University. I don't want to give you the wrong impression … he did excel in math and sciences. I guess he just didn't see the importance of those other classes.

When Gus entered high school he was 5 feet 4 inches and weighed about 100 pounds, not quite what the high school coaches were interested in for the athletic teams, but he was well coordinated and one of the most competitive people that I have ever known, and he tried harder.

Right after high school he went into the Air Force. Shortly after World War II ended he entered Purdue University where he earned a degree in engineering. He then returned to the Air Force and went on to fly 100 combat missions in Korea, became a test pilot, and joined the space program. NASA chose Alan Shepard, John Glenn and Gus as the three astronauts who would be candidates for the first American space flight, ultimately selecting Shepard for the first flight and Gus for the second suborbital flight.

There was a tremendous amount of anxiety in the Grissom house that morning of July 21, 1961, as we all waited for the liftoff of *Liberty Bell 7*. It was quite a relief when we heard that the spacecraft had gone through reentry and had successfully landed. Of course, we later learned that the most dangerous part of the mission occurred in the water, when the hatch unexplainably blew, and Gus almost drowned. The fact that NASA selected Gus as the Command Pilot for the first Gemini flight clearly indicated that they knew that he was not responsible for the hatch prematurely opening.

As a mild extrovert, Gus could surprise you with his wit and humor, and it appeared when you least expected it. He was also a man of few words. He was once asked to speak to the workforce at Convair, a space contractor in Southern California. After a lengthy introduction, Gus got up in front of a couple of thousand workers and gave his famous, three-word speech: "Do good work."

The recovery of *Liberty Bell 7* from the ocean in 1999 exemplifies the pioneering spirit, the dedication and the resourcefulness of Gus. Standing on the dock in that hot July sun, 38 years to the day from liftoff, waiting for *Liberty Bell 7* to be hoisted from the recovery ship, I wondered what Gus would be thinking and feeling as that tiny craft came swinging over onto the dock. I know I had many emotions that were aroused, from deep sadness that Gus wasn't there to see it, to immense pride in knowing that the only craft that he had flown and lost had now come home. Just like it had been said that man could not fly in space, it had also been said that *Liberty Bell 7* was so deep it could never be recovered. Gus was always up for a challenge and I think he would have been very pleased that those who said, "It can't be done," had, again, been proven wrong.

Lowell Grissom, brother of NASA astronaut 'Gus' Grissom, photographed at Grissom Air Force Reserve Base, Indiana. (Photo: U.S. Air Force, taken by Tech. Sgt. Mark R.W. Orders-Woempner, 434th ARW Public Affairs.)

After his Gemini flight, Gus was again selected to be the Command Pilot for the first Apollo flight, leading America to the Moon. Unfortunately, a fire on the launch pad took the life of Gus, Ed White and Roger Chaffee. However, there is a general consensus that America would not have made it to the Moon in the decade of the sixties without the knowledge that was learned, and the corrections that were made as a result of that fire. There is no doubt that Gus would have stepped on the Moon had he lived.

We can honor him only if we follow in his footsteps and peacefully continue to explore space. Our future work in space is bound to include some failures. Yet Apollo 1 has taught us that we can never really fail as long as we persist in our efforts. The greatest lesson we can learn from Gus Grissom is that failure is impossible for those who refuse to abandon their goals. The most fitting tribute to Gus and his Apollo 1 crew is for us to continue doing that for which they gave their lives and to renew our dedication to their quest....REACHING FOR THE STARS!

Lowell Grissom
Mitchell, Indiana
March, 2013

Acknowledgements

This book owes much to those who so willingly assisted in its compilation. In virtually every instance, my request for information or material to quote was met with a positive response.

Sincere thanks are therefore extended to helpers and participants Mike Andrew, John Barteluce, Bob Bell, Rick Boos, Lou Chinal, Dean Conger, Kate Doolan, Lowell Grissom, Donald Harter, Thomas Henderson, Ed Hengeveld, Roger Hiemstra, Jerry Holman (Materium Brush Beryllium and Composites), Philip Kempland, Dale Kreitner, James Lewis, H.H. (Luge) Luetjen, Dr. Robert H. Moser, Lawrence (Larry) McGlynn, Earl Mullins, Otto Preske, Eddie Pugh, Earl Robb, Jerry Roberts, Scott Sacknoff (*Quest* magazine), Ross Smith, Cameron Stark, Charles Tynan, Jr., and Charlie Walker. As well, Dianne Blick, Jim Remar and Shannon Whetzel from the Kansas Cosmosphere and Space Center were of wonderful assistance in locating and supplying rare illustrative material. Other photographs came from two sources, both of whom have – as always – helped out by providing me with clear, high-resolution images; J.L. Pickering of Retro Space Images (*www.retrospaceimages.com*), and Joachim Becker at Spacefacts (*www.spacefacts.de*). Special thanks to the man who located and salvaged the *Liberty Bell 7* spacecraft, Curt Newport, for his interest in – and much appreciated help with – this book.

One of the greatest and long-serving resources of all has been the amazing and immensely popular website, collectSPACE (*www.collectspace.com*), under the inspirational and erudite administration of Robert Pearlman. Robert, and the wonderfully eclectic, knowledgeable band of space enthusiasts who are contributors in many ways to this space website (a daily imperative) have always proved of immense and trustworthy assistance to me, and I am continually grateful to Robert for his dedication and 24/7 input into this website and to the ongoing saga of space exploration.

Keith Scala deserves special mention for contributing so readily to this book. In seeking information for this book I came across an article written by Keith for the quarterly space-flight magazine, *Quest*. Published in the spring of 2000, his article "The Future of Liberty Bell 7" was so well constructed and written that I asked him if he would be willing to

xvi **Acknowledgements**

update the article for inclusion in this book. Happily, he has done so, and I am grateful for his superb contribution.

There were also those who were involved in the editing and production of this book. Endless thanks once again to my esteemed editor for this and past books, David M. Harland. A gifted spaceflight author in his own right, he not only completed a monumental effort in both editing this book and weeding out some pesky errors, but went out of his way and job description to ensure that it went through the production process when problems arose. To a friend of so many decades, Francis French, continuing thanks for going through the draft manuscript as a final check of the facts and my often poor understanding of certain Americanese.

My ongoing appreciation to Clive Horwood and his team from Praxis in England for their support of my past, current and future work, and to Maury Solomon, Editor of Physics and Astronomy, and Assistant Editor Nora Rawn, both at Springer in New York. I recently had the great pleasure of meeting and personally thanking Jim Wilkie, who always provides brilliant cover artwork for my books. He is a genius at what he does.

If I have missed thanking anyone associated with this work, please forgive me, but kindly accept my sincere appreciation for helping me to put together this story of an extraordinary man on an amazing, pioneering space mission in an outstanding Mercury spacecraft he called *Liberty Bell 7*.

Illustrations

Chapter 5

Chapter 6

Appendices

Prologue

At 9:34 a.m. (Eastern Time) on 5 May 1961, the MR-3 combination of a Redstone rocket and a Mercury capsule known as *Freedom 7* lifted off its launch pad at Cape Canaveral, watched by an estimated 45 million viewers across the United States. Onboard, carrying the hopes, prayers, and adoration of a nation was NASA astronaut and Navy Cdr. Alan B. Shepard, Jr. He would successfully complete a suborbital space flight lasting 15 minutes and 22 seconds. In doing so he became the second person after Yuri Gagarin to fly into space, and the first American to achieve that feat.

Two months later a second American astronaut would be seated aboard another spacecraft, ready to fly a similar mission to that of Shepard in order to consolidate the technical data and crucial physiological information gained from that mission. Apart from modifications to this particular Mercury capsule – as recommended by Shepard following his flight – and a far less crowded flight schedule, this second proving flight would follow basically the same test pattern as that *Freedom 7*.

Within those two months between the flights, however, much was happening in regard to America's human space endeavor. Shepard's flight had truly ignited a nation's interest in space flight, and it was now time to capitalize on the success and projected future of NASA's space program, which one day might lead to a human presence on the Moon.

Famously, in his second State of the Union message on 25 May 1961, just 20 days after Shepard's history-making flight, President John F. Kennedy reported to Congress regarding the space program. "With the advice of the Vice President, who is Chairman of the National Space Council," he began, "we have examined where we [the United States] are strong and where we are not … Now is the time to take longer strides – time for a great new American enterprise – time for this Nation to take a clearly leading role in space achievement which in many ways may hold the key to our future on Earth."

In his speech, Kennedy set forth the concept of an accelerated space program based on the long-range national goals of landing a man on the Moon and returning him safely to the Earth; the early development of the Rover nuclear rocket; speeding up the use of Earth satellites for worldwide communications; and providing "at the earliest possible time a satellite system for worldwide weather observation." An additional $549 million in funding was also requested for NASA over the new administration's March budget requests.

At a crowded press conference held following the President's call to Congress, NASA Administrator James E. Webb pointed out to media representatives that the long-range and difficult task of landing a man on the Moon before the end of the decade offered the United States an undeniable chance to overtake and even beat the Soviet Union to this important goal. On 7 June, during an address at George Washington University, a fired-up Webb also stated that the exploration of space was an important part of man's "driving, relentless, insatiable search for new knowledge."

Kennedy's eloquent and challenging speech on 25 May had literally hinged on the success of Alan Shepard's flight less than three weeks earlier, but now he was faced with some serious questions: would Congress embrace and not only agree to what he proposed, but supply the enormous necessary funding? The answer to both questions ultimately rested on the persuasive powers of NASA's highly competitive administrator, who confidently felt the pursuit of funding for the agency's programs was achievable. He had been keeping recent stock of political winds, and realized that Congress – like the rest of the nation – had been swept up in the euphoria and the opportunities offered by human space exploration, and was in what he called "a runaway mood."

Webb's challenge was to come up with the agency's budget forecast figure for placing an American on the Moon by the end of the decade. According to NASA's General Counsel Paul Dembling, the initial projections from Webb's advisors and accountants came in at $10 billion. Dembling was there when Webb scrutinized the numbers. "He said, 'Come on guys, you're doing this on the basis that everything's going to work every time, every place, no matter what you do.' So they came back with a figure of $13 billion."

Once again Webb studied the numbers long and hard before making his way up to Capitol Hill bearing that figure. But when he spoke to the politicians he brazenly stated that the program could cost upwards of $20 billion, and that's what he was requesting. He had applied the old maxim of asking for too much – in this case a whopping $7 billion above the figure his analysts had arrived at – knowing that the enthusiasm of Congress for the space program might soon begin to wane. Webb was right; his ploy worked. He got approval for the money, which would ultimately prove to be very close to the mark by the time the first humans landed on the Moon.

On 22 June, NASA's Deputy Administrator Hugh Dryden sent a letter to Robert S. Kerr, Chairman of the Senate Committee on Aeronautical and Space Sciences, dealing with the broad scientific and technological gains to be achieved in landing a man on the Moon and returning him to the Earth. Dr. Dryden pointed out that this difficult goal "has the highly important role of accelerating the development of space science and technology, motivating the scientists and engineers who are engaged in this effort to move forward with urgency, and integrating their efforts in a way that cannot be accomplished by a disconnected series of research investigations in several fields. It is important to realize, however, that the real values and purposes are not in the mere accomplishment of man setting foot on the Moon but rather in the great cooperative national effort in the development of science and technology which is stimulated by this goal."

Furthermore, Dryden pointed out that "the billions of dollars required in this effort are not spent on the Moon; they are spent in the factories, workshops, and laboratories of our people for salaries, for new materials, and supplies, which in turn represent income for others ... The national enterprise involved in the goal of manned lunar landing and return within this decade is an activity of critical impact on the future of this Nation as an industrial and military power, and as a leader of a free world."

Two days after Dr. Dryden's letter to Robert Kerr, President Kennedy assigned Vice President Lyndon B. Johnson the task of unifying the nation's communications satellite programs, stressing urgency and the "highest priority" for the public interest.

A further two days along, on 26 June, James Webb spoke for NASA in an interview in the *U.S. News and World Report*, stating that "the kind of overall space effort that President Kennedy has recommended ... will put us there [on the Moon] first." This achievement, he said, costing "probably toward the $20 billion level ... will be most valuable in other parts of our economy."

The first salvos in the Space Race to the Moon had been fired. The commitment was there; the money to carry out the activities promised by the President had been made available, and the ambitious plans and goals for American space missions had the overwhelming support of the American people. Certainly the Soviet Union had shot a man into orbit, but the flight of *Freedom 7* with Alan Shepard onboard had enthralled and galvanized a nation. Even though many doubted that the President's stated goal of a man on the Moon could be achieved by the end of the decade, the will to do so was there, while the scientific and technological know-how was in place. The push to the Moon would continue.

To paraphrase the words of Alan Shepard, the first 'baby step' of his brief suborbital flight had amply demonstrated what was required of NASA and the nation's astronauts, and now it was time for America to step up to the plate. There was incredible appeal and an outstanding challenge attached to the task that lay before them.

And so, on 21 July 1961, another of the nation's finest test pilots lined up for his chance at becoming one of NASA's renowned "star voyagers." Strapped snugly into his contour couch aboard a spacecraft he had patriotically named *Liberty Bell 7*, U.S. Air Force Capt. Virgil Ivan ('Gus') Grissom was fully trained and ready to follow in Shepard's pioneering footsteps in order to help to set America on a steady course to the Moon.

1

Creating a Mercury capsule

On Sunday, 16 July 1939, noted scientist Albert Einstein famously sent a letter to President Franklin D. Roosevelt, urging him to explore nuclear weaponry and, as a result, established the United States on the road to the creation of the first atomic weapons ever used to devastating effect in a military conflict.

That very same day an industrial giant was also created when the McDonnell Aircraft Corporation was founded by James Smith McDonnell. Based in St. Louis, Missouri with a startup work force of just thirteen, including McDonnell, it eventually became a leading American aerospace company best known for developing and building some of the finest and most potent fighter jets ever to take to the skies, including the legendary and long-serving F4 Phantom. To those early workers, the McDonnell Aircraft Corporation became more simply known to them by the acronym MAC, and its founder – understandably, and fondly – as Mr. Mac.

Many years later, when McDonnell Douglas merged with the Boeing Company, a new advertising motto was adopted: "Forever New Frontiers." Those three words not only envisaged an exciting future in aviation, but reflected back most appropriately to the glory days of the McDonnell organization.

WITH EYES TO THE FUTURE

James McDonnell was someone always looking to the formidable challenges presented by the new frontier of space, as related by former MAC employee Hulen H. ('Luge') Luetjen.

C. Burgess, *Liberty Bell 7: The Suborbital Mercury Flight of Virgil I. Grissom*, Springer Praxis Books, DOI 10.1007/978-3-319-04391-3_1, © Springer International Publishing Switzerland 2014

In September 1962 President John F. Kennedy visited the McDonnell Aircraft plant in St. Louis. He is flanked in this photo by James S. McDonnell (left) and Sanford N. McDonnell, who became chairman of McDonnell Douglas following the death of his uncle in 1980. (Photo: St. Louis Post-Dispatch staff photographer)

"Mr. Mac had noted, with passionate interest, what had thus far been done in the space arena and announced that in addition to being the world's number one producer of fighter aircraft … McDonnell would also become the world's number one producer of spacecraft – manned spacecraft. He correctly foresaw manned orbital vehicles as being 'just around the corner.'"[1]

In making a commencement speech to engineering graduates at the Missouri School of Mines and Metallurgy on 26 May 1957, some five months before the Soviet Union launched the first *Sputnik* satellite into orbit, McDonnell thoughtfully outlined his expectations for the future of space travel, even giving the students a speculative timetable. Like many others, even though he may have believed that human flight was 'just around the corner,' he did not foresee the explosion of interest in human space flight that *Sputnik* would usher in soon after, and he spoke about the possibility of manned spacecraft orbiting the Earth by 1990. He further predicted that a further 20 years would elapse before there would be a human-tended flight to land on the Moon and return, in about 2010. McDonnell did, however, speak about the escalating threats associated with Cold War tensions, sharing his belief that the United States should instead "wage peace" through the development of dual-use technologies.

"When a chemical rocket motor is developed for a missile, here is a means of propulsion that may be applied in whole or in part to a space vehicle," he told the graduates. "And, when ways are found for a fighter pilot to survive high gravitational pulls at hypersonic speeds, this will help some future space pilot survive blastoff in a Moon-bound rocket."[2] With this futuristic vision firmly entrenched in his mind, McDonnell had even awarded it the code name of Project 7969.

Early in 1958, following the successful launch and orbiting of the Soviet *Sputnik* and the massively unsettling impact this achievement had on the American psyche, McDonnell was more eager than ever to explore the possibilities associated with space travel. A substantial start was made when he established a new department similar to the company's previously established Advanced Design Department (Aircraft), to be headed by L. Michael Weeks, a native of Iowa who had been working on Project 7969 since 1956. Weeks had begun his career teaching mathematics at Iowa State University for three dollars a day before receiving his bachelor's degree in civil engineering at the university in 1943. He had then gone to work with McDonnell Aircraft in St. Louis, eventually rising to the position of chief engineer. In his time with McDonnell he enjoyed key roles in Projects Mercury and Gemini and would also work on Project Apollo and the Space Shuttle. He was later involved with Rockwell International's National Aerospace Plane (X-30) and the Orbital Sciences Corporation's X-34 before retiring after a career spanning 56 years.

The charter for Weeks's department was highly innovative; it was charged with designing a spacecraft capable of carrying a person through launch and into Earth orbit; sustaining that person in space; safely reentering the atmosphere; landing in the ocean, and remaining afloat until the vehicle could be retrieved.

"Ray Pepping, previously Aircraft Chief of Dynamics, became Weeks's assistant," Luetjen recalled, "and John Yardley was named Project Engineer reporting to Weeks and Pepping."[3]

John F. Yardley was a veteran of World War II who had completed his undergraduate education in aeronautical engineering, also at Iowa State University. After receiving his master's degree from Washington University he began his professional career as a stress analyst with McDonnell in 1946. Like Weeks, he would enjoy a long and distinguished career in space flight program development with McDonnell, apart from the years 1974 to 1981, when he joined NASA as the agency's associate administrator in charge of manned space flight. He then rejoined what was by then McDonnell Douglas, serving from 1988 as senior vice president of the merged company.

In March 1958, 'Luge' Luetjen was assigned as Supervisor of Technical Integration under John Yardley. "We knew that studies in many disciplines (aerodynamics, thermodynamics, propulsion, structures, electronics, electrical, design, etc.) would be required," he observed, "and it was my job to keep all of the disciplines 'headed down the same path' and 'singing from the same sheet of music.' As I recall, about 50 to 60

A later photograph of John Yardley. (Photo: Washington University)

full-time people were assigned to the department in short order, with another 20 or so available to be used on a part-time basis as required. Those assigned were the very top people in the various disciplines. What Mr. Mac wanted, Mr. Mac got! Now all we had to do was produce."[4]

Ultimately, James McDonnell's concept of dual-use technology would play a significant role in his company being awarded a contract to build America's first spacecraft; one intended for human space travel and Earth orbit.

THE CONCEPTS OF MAX FAGET

Subsequent to the inception of America's man-in-space programs, Maxime Allan Faget was proving to be a key figure in preparing for this bold new venture, which eventually led to his appointment on 5 November 1958 as Chief of the Flight Systems Division of the newly formed National Aeronautics and Space Administration (NASA).

Faget had attended secondary schools in San Francisco and later trained in mechanical engineering at San Francisco Junior College. In 1943 he received his bachelor's degree in mechanical engineering from Louisiana State University. Following graduation he spent three years in the U.S. Navy, serving aboard submarines for the remainder of World War II.

Post-war, Faget and his former college roommate Guy Thibodaux decided to seek employment together, which led them in 1946 to contact another university friend named Paul Purser, then working at the Langley Aeronautical Laboratory in Hampton, Virginia, which was part of the National Advisory Committee on Aeronautics (NACA). This was the forerunner of NASA, then based at Langley Field in Virginia. NACA, founded in 1915, was a civilian agency dedicated to aeronautical research and development.

Employed as research scientists by Purser, Faget and Thibodaut were first assigned to Langley's Applied Materials and Physics Division working on rocket propulsion, and were then transferred to the Pilotless Aircraft Research Division (PARD). Here, working under division chief Dr. Robert R. Gilruth, Faget was involved in developing engineering concepts on several projects, including the design of a complete ramjet flight test vehicle. He was also a member of the preliminary design team for the hypersonic X-15 research aircraft. Through his prolific talent and determination he was quickly advanced to head the Performance Development branch, where he conceived of and proposed the development of the one-man spacecraft that would ultimately become the Mercury capsule.

Like Faget, design engineer Caldwell C. Johnson from Langley's Technical Services Department enjoyed building elaborately constructed model aircraft – a skill which had been instrumental in landing him the job at NACA straight out of high school. His technical acumen and drawing skills later translated Faget's ideas into working machines. There had been considerable debate in 1956 and 1957 as to whether the United States should attempt to advance the X series of rocket planes in order to carry pilots into space, or whether flying in space would require an entirely new concept. During their lunch breaks Faget, Thibodaux and Caldwell would discuss this at length with others at Langley, and they soon formulated the idea of placing a pilot into an enlarged nose cone atop a rocket and launching him on a ballistic trajectory. No one could find a reason why this would not work if a functional parachute system could be developed, as well as braking rockets to bring the spacecraft back through the atmosphere. It was only a concept, and Johnson sketched out a few prospective nose cone capsules, but it never got much further than idle chatter among some enthusiastic propulsion and design engineers.

Maxime Faget with a model of the Mercury spacecraft. (Photo: NASA)

SPACE TASK GROUP

On 5 November 1958, NASA's Space Task Group, or STG, was created, reporting directly to the Director of Space Flight Development at NASA Headquarters in Washington, D.C. With Robert Gilruth at its head, the STG originally comprised of 27 engineers from the Langley Research Center and another 10 from the Lewis Research Center, plus eight secretaries and "computers." The latter designation was applied to women who ran calculations on mechanical adding machines. They all served as the nucleus for the work carried out on Project Mercury.

As the head of the STG, Gilruth was responsible for reporting to Dr. Abraham ('Abe') Silverstein, NASA's Director of Space Flight Development, who in turn reported to the agency's Administrator, Dr. T. Keith Glennan. The STG included Charles Donlan (Gilruth's deputy); Chuck Mathews (head of flight operations); Chris Kraft (also in flight operations); and Glynn Lunney, who at age 21 was the youngest member of the group. The head of the public affairs office was Lt. Col. John ('Shorty') Powers.

Dr. Robert R. Gilruth. (Photo: NASA)

Work had already begun on the writing of detailed specifications for a Mercury capsule even while the group was still designated as the NACA. By the end of October 1958 a preliminary draft had been completed.

On 17 December 1958 NASA issued an official statement in which the space agency announced the formation of Project Mercury and outlined the program's objectives:

1. To put a manned space capsule into orbital flight around the Earth.
2. To recover successfully the capsule and its occupant.
3. To investigate the capabilities of man in this new environment.

Flight Plan

1. An intercontinental ballistic missile rocket booster will launch the manned capsule into orbit.
2. A nearly circular orbit will be established at an altitude of roughly 100 to 150 statute miles to permit a 24-hour satellite lifetime.
3. Descent from orbit will be initiated by the application of retro-thrust rockets incorporated in the capsule system.
4. Parachutes, incorporated in the capsule system, will be used after the vehicle has been slowed down by aerodynamic drag.
5. Recovery on either land or water will be possible.

Description of Manned Capsule System

1. Vehicle. The manned capsule will have high aerodynamic drag, and will be statically stable over the Mach number range corresponding to flight within the atmosphere. The capsule, which will be of the nonlifting type, will be designed to withstand any known combination of acceleration, heat loads, and aerodynamic forces that might occur during boost or reentry. It will have an extremely blunt leading face covered with a heat shield.
2. Life Support System. A couch, fitted into the capsule, will safely support the pilot during acceleration. Pressure, temperature, and composition of the atmosphere in the capsule will be maintained within allowable limits for human environment. Food and water will be provided.
3. Attitude Control System. A closed loop control system, consisting of an attitude sensor with reaction controls, will be incorporated in the capsule. The reaction controls will maintain the vehicle in a specified orbital attitude, and will establish the proper angle for retro-firing, reentry, or an abort maneuver. The pilot will have the option of manual or automatic control during orbital flight. During manual control, optical displays will permit the pilot to see portions of the Earth and sky. These displays will enable the pilot to position the capsule to the desired orbital attitude.
4. Retrograde System. A system will be provided to supply sufficient impulse to permit atmospheric entry in less than one half an orbital revolution after application of the retro-rockets. These rockets will be fired upon a signal initiated either by a command link from ground control or by the man himself. The impact area can be predetermined because of this control over the capsule's point of reentry into the atmosphere.
5. Recovery System. As the capsule reenters the Earth's atmosphere and slows to a speed approximately that of sound, a drogue parachute will open to stabilize the vehicle. At this time, radar chaff will be released to pinpoint the capsule's location. When the velocity of the capsule decreases to a predetermined rate, a landing parachute opens. The parachute will open at an altitude high enough to permit a safe landing on land or water. (The capsule will be buoyant and stable in water.) After landing, recovery aids will include: tracking beacons, a high-intensity flashing light system, a two-way voice radio, SOFAR [Sound Fixing and Radar] bombs and dye markers.

6. <u>Escape System</u>. In an emergency situation before orbital altitude is reached, escape systems will separate the capsule from the booster. After the capsule is in orbit, the space pilot can reenter the atmosphere at any time by activating the retro-rockets. Other safety control features will be incorporated.

Guidance and Tracking

Ground based and booster equipment will guide the capsule into the desired orbit. Ground and capsule equipment will then determine the vehicle's orbital path throughout its flight. The equipment will be used to initiate the vehicle's descent at the proper time and will predict the impact area.

Communications

Provisions will be made for two-way communications between the pilot and ground stations during the flight. Equipment will include a two-way voice radio, a receiver for commands from the ground, telemetry equipment for transmission of data from the capsule to ground stations, and a radio tracking beacon. This communications equipment is supplemented by the special recovery aids.

Instrumentation

1. Medical instrumentation to evaluate the pilot's reaction to space flight.
2. Instrumentation to measure and monitor the internal and external capsule environment, and to make scientific observations. Note: Data will be recorded in flight and telemetered to ground recorders.

Test Program

As in the case of new research aircraft, orbital flight of the manned space capsule will take place only after the logical buildup of vehicle capabilities and scientific data. Project Mercury includes ground testing, development and qualification flight testing, and pilot training.

A SPACE CAPSULE EVOLVES

Max Faget was one of the original STG formation team. As head of engineering he would personally contribute to the rapid advancement of that program by inventing an emergency escape tower to be used on Mercury and (later) Apollo spacecraft; a 'survival couch' which helped astronauts withstand the accelerations of launch and reentry; and by designing the final configuration of the Mercury capsule interior. However he will always be best remembered for designing the Mercury spacecraft with its iconic blunted leading face (the heat shield area), corrugated sides, and a top end that had the appearance of a screw-on bottle cap. Overall, it looked like an old-fashioned television tube.

Two early design models for the Mercury capsule. On the left is Shape A and on the right is Shape B, with the position of the astronaut indicated in both cases. Before the configuration was finalized, Shape B depicted a proposal very close to the design selected for the craft. (Photos: NASA)

In working with that basic shape, and harking back to his earlier conversations with Johnson and Thibodaux, Faget and his team solved one of the trickiest problems involved in the safe recovery of a manned spacecraft – protecting the vessel and its occupant from the ferocious buildup of heat during reentry. Rather than finessing the streamlined, low-drag shapes that earlier missile nosecones had utilized, Faget conceived of a bell-shaped spacecraft that during a reentry of around 17,000 miles an hour would form a supersonic shock wave well ahead of the blunt, curved heat shield in order to cause a great portion of the aerodynamic heating to occur before reaching the spacecraft.

Asked why a space vehicle should not be aerodynamic, Faget once responded, "Why? Because the higher drag vehicles have less heating during entry than the low drag vehicles. When you enter the atmosphere, when something enters the atmosphere, it slows down on account of drag. Now when you have a blunt face like that you create a huge shock wave, and all the drag is related to the shock wave and all the heat goes into the shock wave. If you don't have that, you've got a very streamlined vehicle. Then you end up with what's normally termed – which is not an accurate term – but it's called friction drag. This drag is taken by the skin friction of the vehicle and all of the heat goes into the vehicle as opposed to it going into the shock wave."[5]

Robert Gilruth stated that the major consideration had always been the shape of the spacecraft and Max Faget was undoubtedly the major contributor, although he recognized that Harvey Allen of the Ames Aeronautical Laboratory was the first, to his recollection, to propose a blunt body for flying a man into space. In Gilruth's words, "In March 1958 Max Faget presented a paper that was to be a milestone in spacecraft design. His paper proposed a simple blunt body design that would reenter the atmosphere

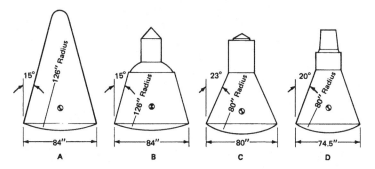

A 1958 sketch of four shapes tested in the evolution of the Mercury capsule. (Illustration: NASA)

without reaching heating rates or accelerations that would be dangerous to man. He showed that small retro-rockets were adequate to initiate reentry from orbit. He suggested the use of parachutes for final descent, and small attitude jets for controlling the capsule in orbit during retro-fire and reentry."[6]

As Faget recalled, the other problem concerning him and his fellow designers was the impact high gravitational forces (g-forces) might have on an astronaut within the space capsule.

"It's quite obvious now that when you launched a man, you put him in a couch so that the Gs come from his back, and then when he reenters, you turn the vehicle around so that the Gs come still from his back. But this was something no one had thought about: how to handle the Gs both during launch and entry. At least they hadn't thought about it very well. I know one of these things, I think it was the Air Force configuration, had studied it enough to decide that, 'Yes, we'd better do something about it,' so they put the man in a sphere and gimbaled the sphere [inside a blunt-nosed capsule similar to one of their missile warheads] so that the vehicle would always be going in the same direction, and they'd turn the man 180 degrees within the sphere so that he could withstand the Gs during entry. [However,] it was ever so much simpler, and the configuration became so much better, if you let the blunt end be the rear during the launch, which would decrease the drag on the launch vehicle, and have the blunt end be forward during entry, where you wanted the drag."[7]

In his role as Chief of the Flight Systems Division at NASA, Faget contributed many of the original design concepts embodied in the Project Mercury spacecraft, and was responsible for numerous innovative spacecraft systems and the task of systems integration.

A CONTRACT IS AWARDED

Meanwhile, at McDonnell's Advanced Design Department, Luge Luetjen said everyone had "hit the ground running" and the place was a beehive of frantic activity.

"The dynamics and aerodynamics people had developed the equations for the ascent trajectory, the orbital flight, and descent trajectory of a typical space vehicle. Working with the engineers in the computer lab, we managed to set up a program such that with any combination of the independent variables (thrust, weight, flight path angle, etc.) the characteristics of the flight could be determined, printed out, and plotted. The results obtained were critical for the structural, heat protection, and aerodynamic design of a spacecraft. Considerable research had been done by Max Faget and others at the NACA Langley Field facility on various spacecraft body shapes and their characteristics. They shared this information freely with all that were interested, and some of us spent considerable time in conference with them on the subject. The 'wheels' in the department, but mostly Yardley, had decided that we would bundle all our studies and calculations in a single report whose format was similar to what we thought that of a Request for Proposal (RFP) for a manned space vehicle might be. As it turned out, we didn't have long to wait."[8]

Even while Robert Gilruth's STG team was still designated as part of NACA, it began work on the writing of detailed specifications for a Mercury capsule. By the end of October 1958 a preliminary draft had been completed.

On 7 November, two days after the official formation of the STG, a briefing of potential bidders for the contract to develop and construct a manned spacecraft was held at the Langley Research Center, where the STG was initially based.

Of the 40 companies in attendance, 20 later indicated that they were prepared to bid for the contract, and were given the preliminary specifications. A week later, on 14 November, NASA had received firm offers from all 20 companies – including McDonnell – to bid on the project. Three days after that, the final documentation, Specification No. S-6 "Specifications for Manned Spacecraft Capsule," was mailed out to the interested parties. The deadline for the return of their proposals was set at 11 December.

"Our report was pretty much on target," according to Luetjen, "and we had little trouble transforming our information into a proposal required to be submitted by December 11. We had our proposal quickly completed and simply spent the remainder of the time double checking and dotting the i's and crossing the t's."[9]

Of the 20 companies that expressed further interest in submitting bid proposals, only 11 actually followed through. In turn, NASA passed these bids on to the STG for assessment. The people at McDonnell were delighted subsequently to learn that when the contender numbers were narrowed by almost half, their company was still in the running. After all their hard work in putting the proposal together, several employees took accrued vacation time in order to be ready to resume work if (or more optimistically, once) they learned their bid had been successful.

Over the Christmas break the STG conducted a scrupulous evaluation of all the proposals and finally narrowed their choice down to the one contractor they felt best qualified from the standpoint of technical abilities, ideas, and approach to the issue. Over in Washington, D.C., meanwhile, NASA officials were also scrutinizing the proposals, conducting an evaluation of the business and management aspects of the twelve bidders.

In January 1959, after Administrator Keith Glennan had reviewed the NASA evaluation, the space agency opened further negotiations with McDonnell as the potential prime contractor. On 12 January, shortly after the McDonnell workers from the Advanced Design Department (Space) had returned from their Christmas and New Year break, all of the department's workers were called to a meeting in Mike Weeks' office where he told them that James McDonnell had been informed by NASA's Space Task Group that following an evaluation of all the submitted proposals, McDonnell Aircraft had been named as the winner of the contract and negotiations would begin immediately for the design, production and support of twelve Mercury spacecraft.

"No one in Advanced Design slept much on Monday night after Mike Weeks' announcement of our winning the Mercury competition," Luetjen recalled. "Each individual was trying to determine what the next move in their expertise should be. As I remember, early on Tuesday, Yardley got the group together and indicated that most, though not all, of those who had worked on the proposal would be joining him on the project. Some of the pure scientists would remain in Advanced Design to do spade work on projects beyond Mercury.

"The early task facing each of the prime systems engineers was two-fold: to work with their NASA counterpart to roughly define their system such that specifications could be drafted on which subcontractors/suppliers could bid; and secondly, to estimate the effort and material costs to consummate their part of the program such that a basic contract could be agreed upon at an early date. It greatly helped that much coordination and contact with potential suppliers had taken place prior to the submission of our proposal. Two main systems were unsettled when it came time to sign the contract, and the contract wording had to be sufficiently loose to allow alternatives. It was yet to be determined whether the heat protection system should consist of a beryllium heat sink or a shield made of a material that would ablate and thus dissipate the heat upon reentry, and secondly, it was unknown at that juncture whether the escape system should be a rocket boost system to separate the spacecraft from the launch vehicle or a rocket pull system in which the escape rocket would be mounted on a tower on the forward (small) end of the capsule."[10]

According to Max Faget, the Atlas rocket had been chosen as the most suitable booster for the later orbital missions. "Bob Gilruth came in one day and says, 'Max … what are you going to do if the Atlas blows up on the way up?' And I didn't have an answer for that. And he said, 'Well, you'd better get an answer for it.' I've always said that was an invention on command. It was very fortunate that one of our colleagues, Woody Blanchard, again in the Pilotless Air Research Division, had been experimenting with tow rockets. He put canted nozzles on a rocket and towed models, research models, up to Mach 1 and above, but instead of pushing, he'd pull them. So knowing that you could do [that with] the rocket up front, it was just a small step instead of putting a cable up there, to put a structure up there that would hold it rigidly during launch but be in place whenever you need it. That turned out to be a very successful thing."[11]

An initial contract for the construction of 12 similarly constructed spacecraft was formally signed on 6 February. This number was subsequently increased to 18, and then to 26, before finally being set to 20 as the development effort matured. During the

Robert Gilruth (left) views an engineering diagram of the Mercury capsule, along with D. Brainerd Holmes, Director of NASA's Office of Manned Space Flight; Walter Williams, Operations Director; and John ('Shorty') Powers, Public Affairs Officer. (Photo: NASA)

negotiations, officials from McDonnell estimated it would be possible to deliver the first three capsules within ten months.

As Robert Gilruth noted in his unpublished memoirs, "During this same period of time we established an arrangement with the Ballistic Missile Division of the Air Force for the procurement of the Atlas launch rockets and for launch services. We [also] worked out a plan with [Major] General [John B.] Medaris [commanding the Army Ballistic Missiles Agency] and Dr. [Wernher] von Braun [of that agency] for the Redstone launch vehicles, and we started work in our own staff for a design and specification for the Little Joe rocket to be used in tests at Wallops Island. We gave to Lewis the job of creating a full-scale Mercury model spacecraft for an unmanned flight at an early date to establish levels of heat transfer and stability in a full-scale free-flight test on an Atlas booster at Cape Canaveral …. The project was started in December 1958 and flew successfully in September 1959."[12]

DEVELOPMENT OF THE CAPSULE

The contract signed with McDonnell Aircraft was a cost-plus-fee type, with a cost figure of $18.3 million dollars and a fee of $1.15 million dollars. As Luge Luetjen indicated, the contract "set in motion what was to eventually become one of the largest technical mobilizations in American history involving some 4,000 suppliers, nearly 600 direct contractors from 25 states, and over 1,500 second tier subcontractors. To manage such an effort, Mr. Mac and Dave Lewis (by then second in command) named Logan MacMillan, a former chief test pilot and head of flight test, as the Mercury Company-wide Program Manager. Under Logan was E.M. ('Bud') Flesh as Engineering Manager, Bill DuBusker as Manufacturing Manager and, of course, [John] Yardley as Project Engineer. After a short time, NASA felt there should be someone at vice presidential level in charge, so Mr. Mac designated Walter Burke, already a vice president in charge of the entire factory, as the Vice President and General Manager of the Mercury program."[13]

Beginning in February 1959, endless tests of different small-scale spacecraft shapes began in wind tunnels across the United States. Engineers, technicians and designers then studied the resulting flow effect in photographs, looking closely at phenomena such as shock waves, wind streams, and vortices. By the end of the year some 70 design models had been tested and analyzed, with results proving a full-scale capsule could handle the tremendous aerodynamic stresses associated with launch and reentry. As well, they established that the blunt shape at the bottom of the capsule would provide sufficient drag to slow the craft down during its descent through the atmosphere. In tests, NASA engineers applied 6,000-degree jet flames onto that same blunt end to demonstrate its ability to withstand the incredibly high temperatures associated with reentry.

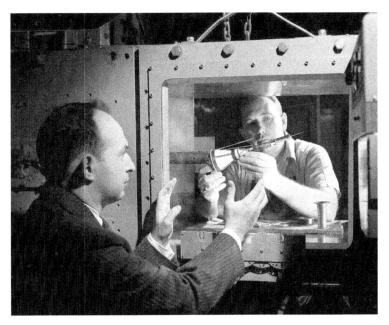

Testing a small-scale model of the Mercury capsule/escape tower assembly in an 18 inch by 18 inch wind tunnel. (Photo: NASA/Langley Research Center)

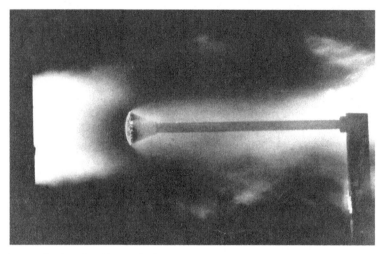

Scientists at the Langley Research Center operated hot jets, quartz-lamp heat radiators, furnaces and other specialized research facilities in making tests at temperatures up to 10,000°F on space vehicle structures and materials. This scale model, made of fiberglass and plastic, was exposed to a 5,000°F arc-heated air jet. (Photo: NASA/Langley Research Center)

A Mercury capsule prototype undergoes testing in the Full Scale Wind Tunnel at Langley Research Center in January 1959. (Photo: NASA/Langley Research Center)

Once all the preliminary design work was complete, McDonnell manufactured several scale models of their own for testing purposes. As Max Faget, Caldwell Johnson and others were no longer needed in the design development of the capsule, the MAC team also went to work on testing full-scale, non-functional "boilerplate" models.

The McDonnell Mercury Design Team (circa 1959). From left: Jack Crouch, Chuck Jahn, Earl Younger, Norris LaGrant, Bob Roth, John Dale, 'Bud' Flesh, John Yardley, Bill Mosley, George Weber, Harry Condit and 'Luge' Luetjen. (Photo: McDonnell Aircraft Corporation)

To bridge the gap between the signing of contracts to construct the spacecraft and their actual delivery, and to aid in astronaut training, the Langley Research Center employed full-size replica Mercury capsules constructed from steel plates. These became known as "boilerplate" mockups. Test launches of boilerplate models were carried out using Sergeant and Recruit rockets, while others were conducted at Cape Canaveral using Redstone, Atlas and Scout boosters.

OF PIGS AND ASTRONAUT COUCHES

"Those were the days of the most intensive and dedicated work of a group of people that I have ever experienced," Gilruth once wrote in a conference paper. "None of us will ever forget it. We were making tests of escape rockets over on the beach at Wallops

Island, testing parachutes in full-scale drops from helicopters, and measuring water impact loads on capsule configurations at Langley Field."[14]

Meanwhile, impact studies were also being carried out at the McDonnell plant, with test capsules dropped into water, onto sand beds, and onto solid concrete slabs. In one series of tests, live pigs were loaded into a Mercury capsule, strapped into a specially built contour couch similar to those being individually developed for the astronauts. The capsule was then dropped down a long shaft blunt end first to verify that a human passenger could survive a particularly hard landing, as described in the official NASA Project Mercury history.

"Through April and May (1959), McDonnell engineers fitted a series of four Yorkshire pigs into contour couches for impact landing tests of the crushable aluminum honeycomb energy-absorption system. These supine swine sustained acceleration peaks from 38 to 58 g before minor internal injuries were noted. The 'pig drop' tests were quite impressive, both to McDonnell employees who left their desks and lathes to watch them and to STG engineers who studied the documentary movies. But, still more significant, seeing the pigs get up and walk away from their forced fall and stunning impact vastly increased the confidence of the newly chosen astronauts that they could do the same. The McDonnell report on these experiments concluded, 'Since neither the acceleration rates nor shock pulse amplitudes applied to the specimens resulted in permanent or disabling damage, the honeycomb energy absorption system of these experiments is considered suitable for controlling the landing shock applied to the Mercury capsule pilot.'"[15]

Jerry Roberts was a Guidance and Control System Engineer working at the Cape with McDonnell, and he still remembers conducting those drop tests using live animals.

"When we were designing the spacecraft we knew nothing about the effects of weightlessness on the astronaut – on the human body. We also knew very little about what would actually happen during the launch and recovery process. We knew it would be pretty rugged, so one of the things we had to do was design a seat for the astronauts to give them the maximum protection possible in this confined space. And believe me it was confined; the men could just barely fit into the Mercury spacecraft.

"The small MAC group that I was with came up with a couch for the astronauts to sit in, molded around the astronauts' bodies. It was constructed out of a honeycomb material. We did this individually for each astronaut." The couches, backed by an energy-absorbing, crushable aluminum honeycomb, consisted of a fiberglass shell with protective rubber padding, and were located in the pressurized section of the capsule.

"Then we needed to test it to see how much protection it actually provided, and since most of the astronauts weighed about 160–180 pounds, someone came up with the idea of using pigs that were in that weight range.

"We had a test facility constructed inside one of our lab buildings and it was fixed so that we could take these hogs up to varying heights. They were sedated of course, and we dropped them down a chute into a bed of wet, packed sand. We recorded the Gs that each one was subjected to, but I know we started out at about eight feet and then went [up] in four foot intervals. I think the last pig was dropped at 20 feet, or

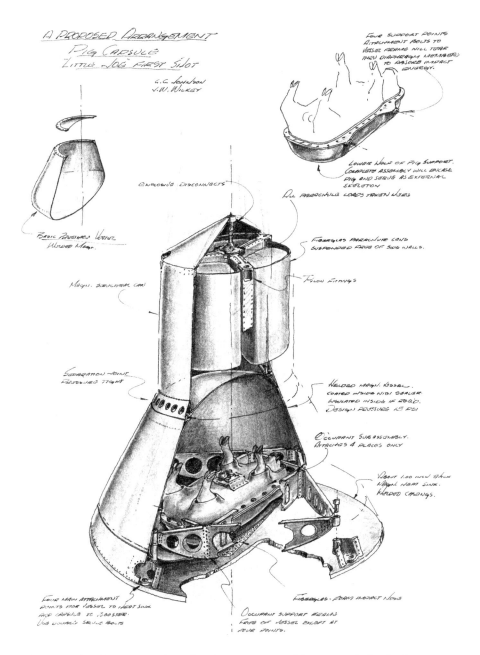

This illustration demonstrates how the pigs were strapped into a couch for the drop tests. (Photo: NASA)

maybe even higher. As it turned out the seat did provide significant protection and so the test was successful.

"Immediately after each pig was subjected to the drop test the animal was butchered; the local butcher did all the slaughtering right there in the facility and the pig's internal organs were examined in great detail to see what damage resulted from the fall. The meat was frozen and I think was later donated to [some charity] the equivalent of a soup kitchen. It was not thrown out or wasted."[16]

PROTECTING THE ASTRONAUTS

After much deliberation and testing, a decision had been reached on the style and composition of the capsule's heat shield. For the initial suborbital flights, it had been decided to adopt a proven system known as a 'heat sink,' which had been developed for the ballistic missile program. Previous testing had revealed that although the intense shock wave generated by a missile cone's trajectory through the atmosphere managed to keep the massively high temperatures away from the forward-facing blunt end of the cone, enough heat – estimated at a temperature of around 3,000°F – could potentially soak through to melt or even vaporize in an explosive release of gases any normal metal, greatly endangering the life of an astronaut. However, beryllium, with its unusual ability to absorb extremely large quantities of heat, was the obvious candidate to test as the heat sink for a manned capsule.

On Monday, 8 June 1959, after details had been kept secret to that time, it was announced that the Brush Beryllium Company, which operated a plant near Elmore, Ohio, had been assigned the task of producing six gently curved heat shields to protect astronauts from the tremendous frictional heat encountered when their spacecraft reentered the atmosphere.

Beryllium is a hard, light metal that has a high melting point and it was used due to its ability to absorb heat as well as its high conductivity, preventing disastrous build-ups of concentrated surface temperatures. Specifications called for the heat shield to be constructed of "hot-pressed" beryllium, with a diameter of 80 inches and a radius of curvature of 120 inches. It would prove to be the largest single piece of beryllium ever forged to that time.[17]

In July 1959 Brush Beryllium and the Aluminum Company of America announced the successful production of the first giant, dish-shaped beryllium piece, forged by Alcoa from a record-size billet supplied by Brush.

To produce the heat-sink shield, Brush first hot-pressed a beryllium billet 62 inches in diameter, one of the largest ever made to that time using powder-metallurgy techniques. This was achieved using the company's patented QMV (quantum mechanical vacuum) process, involving simultaneous applications of vacuum, heat and pressure to beryllium powder. Following preliminary machining by Brush, the billet was encased in steel for the high-temperature forging operation. It was then delivered to the Alcoa factory in Cleveland, where it was heated to approximately 2000°F in a specially designed furnace. A huge manipulator then removed the glowing, steel-jacketed beryllium piece and placed it onto a pre-heated die. The mighty force of a 50,000-ton press, operated by Alcoa under the U.S. Air Force's Heavy Press Program, squeezed the beryllium billet into a saucer-shaped disc 80 inches across and three inches thick.

Under the contract, Brush Beryllium then forged the final dimensions in their precision machine shop in Cleveland. The last operation in the manufacturing process – ultrasonic inspection – was carried out by Alcoa. Following this, the McDonnell Aircraft Corporation received the finished piece, 72 inches in diameter, ready to be installed as a heat sink of one of the Mercury spacecraft.[18]

While a beryllium heat shield would be used on capsules in the early booster test flights and the two suborbital missions of Shepard and Grissom, for orbital missions a new, ablative heat shield weighing far less was developed for the Mercury-Atlas flights that would follow.

MERCURY ASTRONAUTS

Seven American test pilots leapt to instant prominence on 9 April 1959, when NASA formally announced their names at a Washington D.C. press conference, introducing them as the space agency's Mercury astronauts.

After several weeks of orientation lectures by members of the STG, each of the seven men had been assigned a specific area of specialization and responsibility to pursue. This came about after NASA realized that the entire scope of Project Mercury was so broad, and areas of development so numerous, that it was almost impossible for all seven astronauts to stay in contact with all the latest developments. Thus, at regular meetings, they would individually report on progress and any problems within their specific assignment. This meant that all seven astronauts were kept up to date on the latest developments without the need for them to be involved in studying or contributing to all areas connected with the Mercury program. These assignments were:

Scott Carpenter – Communications and navigation
Gordon Cooper – Redstone booster
John Glenn – Cockpit layout
Gus Grissom – Electromechanical and autopilot systems
Wally Schirra – Life support systems
Alan Shepard – Tracking and recovery
Deke Slayton – Atlas booster

One aspect of the job in which all seven astronauts played an active part was visiting various contractor facilities in order to familiarize themselves with such things as mockups, hardware, and manufacturing processes. For instance, following his selection as a Mercury astronaut, Marine Lt. Col. John Glenn was assigned the task of working with the McDonnell engineers to help determine the layout of the capsule's instrument panel. Now, with the basic shape of the spacecraft fully established and approved, final design and development work on the cockpit instrumentation could begin.

The seven Mercury astronauts. From left: Wally Schirra, John Glenn, Deke Slayton, Gus Grissom, Alan Shepard, Scott Carpenter and Gordon Cooper. (Photo: NASA)

First of all, as Glenn recounted in *We Seven*, "McDonnell had to figure out a way to build [the capsule] so it would be as strong as possible and as light as possible at the same time. The engineers knew that every pound saved on the pad would provide an additional mile in range."

As he explained, the wall of the capsule was made up of two layers of high-grade metal. "The outer layer consists of shingles made from a metal called René 41. These have been corrugated and then welded together to give them extra strength. The welding technique had to be specially perfected so that the thin sheets of metal would not be torn or cracked in the process. The inner layer is made of titanium, a light, strong metal which was developed for jet engines and provides the strength of steel at about half the weight. The two layers are separated by a hollow space that provides extra insulation. It was an extremely difficult vehicle to build, and it was full of compromise. It was not perfect, but it was functional."[19]

Gus Grissom's prime responsibility was working on the Automatic Flight Control System and autopilot, especially for the upcoming orbital missions.

"The path that the capsule follows [after launch] can't be altered after we come off the Atlas booster. Once we are in orbit, we can't change that orbit. As we rotate around the Earth, the autopilot will maintain us in a position to be always looking at the Earth – which actually means that the capsule has to be turned 360 degrees each time we go around the Earth. If we want to change the position of our capsule and look in another direction, or if the autopilot should malfunction, we can then take over with the Manual Attitude Control System. To fly the Manual System we have a side arm controller; it is very similar to the control stick in an airplane – except that an airplane has rudder pedals also, while in this we have eliminated the rudder pedals and made it a function of the stick also. We have a three-axis control."[20]

MERCURY CAPSULE TAKES SHAPE

"We are building and designing the capsules at the same time," Edward ('Bud') Flesh commented at the time. Flesh was the McDonnell engineer in charge of the project. "The design is not complete until we turn out a capsule, and each capsule will be slightly different from the one before, depending on whether it will be a test model or will carry an animal or an astronaut."[21]

The capsules were assembled in rooms that could have rivaled hospital wards for cleanliness. Technicians and engineers wore white clothing made of dust-free nylon, and shoes of white nylon. The rooms, also white, were air-filtered and temperature-controlled.

McDonnell technicians working on a Mercury capsule, 1960. (Photo: NASA)

Psychologists, physiologists and engineers were all taking part in the design process, according to Fred Willis, one of the project engineers. "It would be silly, for example, to put a red warning light more than 50 degrees to the left or right of the man's line of sight. Only the cones of our eyes see color, and the cones are not there for peripheral vision. A red light at the edge of our vision would not be recognized instantly as a danger signal. Maybe it seems a small point, but attention to detail like this makes the space ship as safe as a living room."

One sticking point in the development of the first Mercury capsules was the subject of fitting viewing windows. Bud Flesh said it was a design problem that they had overcome. "As pilots, the astronauts are used to having a windshield," he pointed out, "and even though it is probably more a psychological than a physical matter, we of course gave it to them."[22]

Max Faget agreed on the subject of a viewing window – it was one of the design changes that had been unanimously demanded by the astronauts. "Oh, there was a big beef about that," Faget told interviewer Jim Slade. "We had two little portholes about that big around; six-inch round quartz windows, one down here and one over here, and it wasn't very good. And another thing we had is, in order to navigate, we had essentially – it was like a periscope, only actually it was optics to show them a virtual image here of the ground passing underneath them. We thought that was very important, that they see the ground so they could line it up with a reticule in there to make sure the vehicle wasn't yawed. If they could see a piece of ground going right down along the center line, well they knew that they were headed right so when they fired the retro[rocket] off, they would be properly aligned. Actually, the [attitude] could be misaligned some fifteen or twenty degrees and it wouldn't make that much difference. It would move the landing point further down range, but that's about all. We had more than enough capacity [to retrieve the capsule]".[23]

Astronaut John Glenn shows his wife Annie a replica of the Mercury spacecraft. (Photo: NASA)

The U.S. Navy human centrifuge at Johnsville, Pennsylvania. The radius of the arm is 50 feet and the gimbal at the end of the arm changes positions as the centrifuge arm turns, by computer or by manual pilot control in the gondola at the end of the arm. The gondola and astronaut may be pressurized or depressurized for altitude as the accelerations are provided. (Photo: U.S. Navy)

All seven astronauts visited the McDonnell plant in May 1959, where each man was assigned an appropriate systems engineer to provide individual presentations on the status of the different systems – their own particular part in them – and acquaint them with the people at McDonnell and the plant's layout. After that they visited the Cape Canaveral launch site and trained at a number of military and medical facilities in accordance with the training program prepared for them by the STG. That August, for example, they were involved in human centrifuge tests at the Naval Air Development Center in Johnsville, Pennsylvania, learning to cope with excessive g-forces and the accelerations associated with their flights into space and back. They were unanimous that the centrifuge, which they referred to as "The Wheel," was the hardest part of their training to that time.

As Alan Shepard explained, "This thing – the centrifuge – puts you in what pilots call the 'eyeballs-out' position. It is like an oversize cream separator that flings you around the room, that sort of thing. If you fight it, it's murder. You can easily black out and remain unconscious. When I come off that monster at Johnsville, every bone in my body aches."

Scott Carpenter agreed with Shepard. "It's important to master these g-forces because we fully expect to be operating the capsule controls during part of the flight. So far, we have all shown this capability under 9 Gs. But that's assuming everything goes according to plan. If it doesn't … well, let's not talk about that."

"Just say it is a real personal challenge," added Shepard.[24]

In September the Mercury astronauts returned to the McDonnell plant to check on progress with the spacecraft that they hoped they would soon fly. As Luge Luetjen recalls, "During this and subsequent visits, we developed lasting first-name friendships and a mutual respect for each other's role in the 'great adventure.' Their suggestions and recommendations throughout the program were timely, well thought out, and were of great help in the design, construction, and testing of the machine."[25]

PARACHUTE SYSTEMS

Another of the problems that needed to be addressed by Langley's Flight Research Division – prior even to the formation of the STG – had been the development of a reliable parachute system for a manned space program. Beginning in October 1958 a progressive series of air tests were carried out to assess the deployment, reliability and specifications of a number of different parachute systems.

The first air drops were conducted as a means of studying the free-fall stability of the spacecraft, parachute shock loads, and the operation of the capsule's escape system. Initial tests were carried out by the Pilotless Aircraft Research Division at the High Speed Flight Station, Edwards AFB, California, in order to collect data on opening characteristics and shock loads associated with the drogue chute. Once this information had been gathered and collated, the test engineers' attention turned to the size and performance characteristics of the main parachute. This had to be large enough so that the final impact velocity of the capsule might be kept at about 30 feet per second.

A D-shape capsule features in this sequence of photographs from a beach abort launch-to-parachute test on 13 April 1959. (Photo: NASA/Langley Research Center)

In order to demonstrate the adequacy of the mechanical system in deploying the drogue and main parachutes, preliminary drops were made from NASA helicopters at West Point, Virginia. These utilized concrete-filled drums attached to the operating canister system. Following these tests, a Lockheed C-130A Hercules cargo aircraft was supplied by the Tactical Air Command for the continuation of tests, now involving both high- altitude and low-altitude drops.

Initially, low-level drops were carried out in the vicinity of Pope AFB, near Fort Bragg in North Carolina, to perfect the best means of extracting a full-scale capsule equipped with operating parachute systems from the open tail ramp of the C-130. Once these had been completed, the research and development program moved on to Wallops Island, Virginia, where further drops were carried out under the auspices of Langley's Flight Research Division. The advanced tests were planned to study the stability of the Mercury capsule both during free flight and with parachute support, shock input into the capsule by the parachute, and retrieval operations. Four drops were completed from altitudes ranging up to 23,000 feet, with parachute openings at up to 15,000 feet. These successful tests demonstrated that with properly designed equipment, there was no impediment to recovery helicopters being utilized in the retrieval operation.

Subsequent air drop tests were completed at various altitudes to investigate the stability of the capsule using a 6-foot FIST (*Flugtechnisches Institut Stuttgart*) Ribbon drogue parachute in combination with a 67-foot extended skirt main chute. These indicated that a different type of main chute would offer greater reliability. Ring-sail parachutes were substituted, and the drop-test program continued. The results of these tests concluded that a ring-sail parachute would have the desired reliability.

Six succeeding drops using a 6-foot FIST drogue and the 63-foot ring-sail main canopy were all successful. It was decided that this parachute combination worked best and could be used throughout the Mercury program.

This blurred, long-distance image taken from film footage of a drop test shows a Mercury boilerplate being jettisoned from a C-130 airplane. (Image: NASA)

Alan Shepard demonstrates what an incredibly tight and difficult squeeze it was for an astronaut to egress through the top of a Mercury capsule in his space suit. (Photo: NASA)

As Max Faget explained, "What happened was, the upper part of that capsule held the parachutes. There were two parachutes, the main parachute and the backup, which were identical – completely identical. Even the systems were all identical. They had their own drogue parachutes and pilot parachutes and everything else, but we never had to use the backup system. And there was a hatch up in front, up at the top. If you were sitting in the capsule on the pad – of course, it's a cone around you like that – right up there would be this hatch, and that was nothing but a dish that was held in place, more or less, by the pressure, although it had a few latches, an inwardly opening door, which made it very light, and, of course, it was dish-shaped, so that it was just about as light as you could make it.

"So when the thing got on the water and [was] floating upright, you could unfasten this dish and push the containers for the parachute, just push them out; they'd fall overboard, and you could get out …. Well, the astronauts did not like the idea of being trapped in this thing, so they complained about it, and we put this explosive bolt device on the side there, which had to be [powered] on and then fired.

"The hatch that they went into was fastened with something like about fifty or sixty small bolts, so it really wasn't a hatch, it was just a covering. So they were essentially sealed in the capsule."[26]

John Glenn demonstrates the egress technique in a test tank. (Photo: NASA)

In fact, as revealed in NASA's configuration specifications to McDonnell for the Mercury capsule, there were actually 70 bolts in the emergency egress hatch:

3.5.3 ENTRANCE AND EMERGENCY EGRESS HATCH – The entrance and emergency egress hatch, in accordance with Drawing No. 45–35003, located in the capsule conical section, shall be trapezoidal in shape as dictated by the capsule configuration. The hatch assembly shall be of a construction similar to the basic capsule structure, designed to permit entry into, and emergency egress from, the capsule. An explosive assembly, in accordance with Drawing No. 45–35701, shall be incorporated in the hatch assembly to serve as a means, when ignited, of breaking the seventy (70) hatch attachment bolts. The explosive assembly shall be mounted about the hatch perimeter and shall consist of a gasket type sill containing a continuous single strand of explosive charge to effect severance of the attachment bolts. The strand shall be ignited from both ends simultaneously to provide redundancy. A push-button initiator, located on the hatch interior to the astronaut's upper right, shall, after removal of a safety cap and pin, ignite the explosive charge when pushed by the astronaut. A pull initiator assembly shall be provided for ground rescue utilization on the exterior of the hatch beneath the shingles. Function of the pull initiator assembly shall be the same as for the astronaut-actuated initiator. The hatch assembly shall be secured to the capsule structure by two wire springs, in accordance with Drawing No. 45–35058. These springs shall absorb the energy expended by the explosive charge and serve to prevent injury to personnel working in the hatch area during recovery operations.[27]

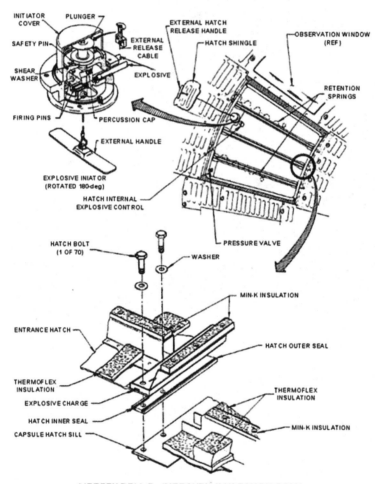

LIBERTY BELL 7 - (MERCURY 4) HATCH DIAGRAM

Diagram of the MR-4 hatch. (Illustration: NASA)

FINAL CONFIGURATION

Stated simply, in its final configuration the bell-shaped Mercury capsule was comprised of a conical pressure section topped by the cylindrical recovery system section. The beryllium heat shield was located at the base of the cone and a retro-rocket housing was held to the shield by three metallic straps.

The three retro-rockets, enclosed in a housing in the center of the heat shield, provided the reverse propulsion required to slow the orbiting spacecraft by about 500 feet per second and thereby initiate the deorbit process. The expended retro-rocket package would then be jettisoned.

Attached to the top of the spacecraft by explosive bolts was the escape system, made up of an escape rocket on a 14-foot tubular tower. The escape system was designed such that the Mercury capsule, in an abort contingency, was explosively disengaged from the booster as the escape rocket fired simultaneously. The rocket would pull the capsule upward and to the side as it separated at a rate exceeding 200 feet per second. Once the capsule/escape tower combination had slowed down, the tower would be jettisoned and a normal parachute sequence initiated.

The spacecraft's instrument panel was set approximately 24 inches in front of the astronaut. The environmental control system provided conditions for the astronaut similar to that of a military aircraft and had redundant controls, plus another one for emergencies.

The recovery system included the 6-foot FIST drogue parachute which opened at around 21,000 feet and the 63-foot ring-sail main parachute that would be deployed at about 10,000 feet, plus the reserve drogue and main chute. As tests indicated, the planned landing velocity in the water was somewhere in the optimum vicinity of 30 feet per second.

MOVING TO THE CAPE

In a McDonnell inter-office memo dated 25 August 1959, Bud Flesh announced that a Project Mercury Operations Group would be established at Cape Canaveral, with the office to become active by 3 September. Flesh stated that the group would be led by Luge Luetjen, Assistant Manager of the Operations Group and Engineer in charge of Technical Integration. His responsibilities included establishing the MAC office, liaising with NASA and other space flight and military authorities, and representing the company on committees set up to oversee Mercury Operations. He would also coordinate Redstone Capsule Launch Procedures with the Missile Firing Laboratory (MFL) and NASA.

Among other responsibilities outlined in the memo, Guenter Wendt was assigned to carry forward all arrangements with the Redstone MFL that were necessary for a coordinated program. In particular, he was to assist Luetjen in the launch procedures coordination and in Redstone missile committees. Wendt had been employed by Luetjen because he needed someone who was "a combination of pseudo-engineer and non-union technician to help in some of the technical areas and still do 'grunt' work when required."[28]

Writing later on the man who would become affectionately known to all at the Cape as the "pad fuehrer,' Luetjen remarked that Wendt had "allegedly" been an air crew member in the German Air Force (Luftwaffe) and had worked as a mechanic for Ozark Air Lines before joining McDonnell.

The MAC contingent arrived at Cocoa Beach on 2 September and settled into rooms at the Satellite Motel located north on Highway A1A – the main street of Cocoa Beach – linking the beach area to Merritt Island and the City of Cocoa. The next morning they proceeded to the south gate of the Cape where they were met by an administrative officer from NASA who handed them their official badges and escorted them to Hangar S, where they would occupy the south balcony in an area recently vacated by the Martin Company people involved in the Vanguard project.

Jerry Roberts was one of those assigned to work at the Cape, and he recalls times when he and his McDonnell co-workers had to work long hours, seven days a week,

Guenter Wendt with Gus Grissom's *Liberty Bell 7* spacecraft. (Photo: NASA)

Hangar S at Cape Canaveral. (Photo: NASA)

often putting in as many as 100 hours per week. "When we first were transferred from St. Louis to the Cape in 1960 it was supposed to be for nine months, and we were going to launch these things initially and then train the NASA people how to launch them, and then we were to come back home. That was pretty evidently not going to happen – NASA had no such capability to even learn from us. They didn't have the people with the background to take over either part of the program – the missile or the spacecraft – so instead of being down there nine months, some of our people were there till the end of the Gemini program [in 1966]."[29]

At the time, Operations Director Walt Williams from the STG spoke admiringly of the spirit prevailing at the Cape. "When you have people so well motivated, which they are, you find them working terribly hard and doing really good work. Of course, there should be physical limits. Our people at Cape Canaveral were working 80 to 90 hours a week, which was just too much for them. So we put an arbitrary 60-hour limit on them. Then I found out they were working anyway – just not getting paid for it. This is the kind of dedication we have all through the program"[30]

Atlas 10D on Launch Pad 14 with the Big Joe boilerplate Mercury capsule. (Photo: NASA)

The Big Joe Mercury-style capsule is shown aboard the USS *Strong* (DD-467) after being recovered from the Atlantic Ocean on 7 September 1959 several hundred miles northwest of the island of Antigua. The top of the capsule held the recovery system during the flight, and was opened to deploy the drogue parachute which slowed the descent of the capsule after reentry and the large main parachute that lowered the capsule to the water. Also visible in the exposed top are the high-intensity marker light and the antenna (folded down) of the search and recovery beacon radios. The small vertical light areas on the side of the capsule are the flush-mounted telemetry transmitter antennas. (Photo: NASA)

A week after their arrival, on 9 September, the MAC team witnessed the vitally important test launch of Big Joe 1 (Atlas 10D) that sent a 2,555-pound, unmanned boilerplate Mercury capsule on a ballistic arc a distance of 1,424 miles with a peak altitude of 90 miles. This particular capsule was not equipped with a launch escape system. The principal objective of the Big Joe program was to test the Mercury spacecraft's ablating heat shield, which would be used on the later orbital manned missions. It was also the first Project Mercury flight to employ an Atlas booster. However the booster failed to stage correctly, and separation from the Mercury boilerplate occurred far too late. The capsule was eventually located and retrieved from the Atlantic Ocean and subsequently studied for the effects of reentry heat and any other flight stresses resulting from the 13-minute flight. Despite the booster malfunction, the heat shield had survived the reentry phase and was found to be in marginally good condition. More work needed to be done.

As Luge Luetjen commented, "The telemetry had functioned properly until [the radio] blackout, and as the morning wore on and the spacecraft was recovered and the data reduced, it became obvious that there was sufficient heat pulse to prove the ablation design if the physical examination results were positive. Upon arrival at the Cape, eager eyes focused on the heat shield and marveled at its superb condition. As a result, all plans to use the heavier heat-sink design were scrapped for the rest of the [Mercury-Redstone] program."[31]

Indeed, data gathered from the Big Joe 1 flight was enough to satisfy NASA that they could cancel a second launch – Big Joe 2 (Atlas 20D) – that had been planned for the fall of 1959. The launch vehicle manifested for this flight was then shifted to another program.

McDonnell would only miss their optimistic 10-month delivery forecast by two months. Mercury Spacecraft No. 4, planned for the MA-1 test flight, was delivered to NASA on 25 January 1960.

Shown here in July 1959 are Mercury-style capsules specially manufactured by Langley technicians for the Little Joe series of proving flights from Wallops Island, Virginia. While these separate designs were not intended to carry astronauts, they would fly monkeys to test many facets of launch and recovery, including center of gravity requirements and aerodynamic loads. (Photo: NASA/Langley Research Center)

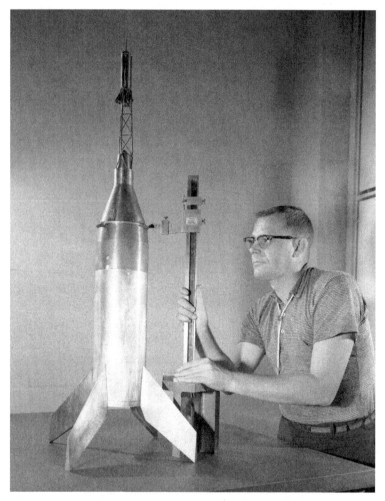

A technician checks a wind-tunnel model of the Little Joe/Mercury capsule combination, circa 1959. (Photo: NASA/Langley Research Center)

ASTRONAUT WATER SURVIVAL TRAINING

To assist in astronaut water survival and egress training, McDonnell supplied NASA with a full-size boilerplate Mercury capsule designated Unit SC-5. The first phase of this crucial training program took place in February 1960 at the Langley Flight Research Center's hydrodynamic basin – an open tank of fresh water 2,200 feet long by just 8 feet wide and 5 feet deep. Each of the astronauts, dressed in regular flight suits, had to make several egresses through the top end of the floating capsule in both still water and simulated waves.

While assembling a Little Joe craft a workman tapes a signed $20 bill to the inside of the capsule as a souvenir of the flight. (Photo: NASA/Langley Research Center)

Navy crewmen maneuver a huge net alongside the salvage ship USS *Preserver* (ARS-8) to retrieve the unmanned Little Joe 1-B capsule five miles off the coast of Wallops Island. (Photo: NASA)

During impact studies conducted in Virginia's Back River, the astronauts practiced the dangerous maneuver of exiting a floating spacecraft after splashdown. (Photo: NASA)

The second phase of this training began two months later in the Gulf of Mexico, offshore from the Pensacola Naval Air Station in Florida. Wearing astronaut-style flight pressure suits, the men now had to practice egress techniques which involved clambering out of their couches and making their way up through the top of the mockup spacecraft. They also had to practice egress methods from the open side hatch, all the while supported by Navy frogmen.

The third phase of this training commenced in August 1960 back at Langley. On this occasion the boilerplate SC-5 was fully submerged in the center's hydrodynamic basin and each astronaut now had to egress the capsule six times – thrice wearing flight suits and thrice again wearing flight pressure suits.

THE FIRST MERCURY-ATLAS SHOT

By July of 1960 there were close to a hundred McDonnell people employed at the Cape, and things were moving ahead rapidly as the first full-configuration Mercury spacecraft was to be delivered there on 24 July. In order to keep all the spacecraft free from dust or other intruding contamination, the vehicles were all delivered to the Cape in clear plastic sheathing and transported to another clean room within Hangar S. This particular capsule was Spacecraft No. 2, destined to fly on the unmanned MR-1 test flight of the Redstone/capsule combination. Prior to this, Spacecraft No. 4, which had been delivered to the Cape on 23 May, was due to ride an Atlas rocket for the MA-1 proving flight.

The MA-1 spacecraft shell had been loaded with around 200 pounds of sensing instrumentation, installed by NASA Langley. As with the earlier Big Joe launch, there

Langley engineers practice water egress techniques using replica capsules that are surrounded by inflated air bags. (Photos: NASA/Langley Research Center)

was no escape tower attached to the capsule, much to the dismay of the Atlas rocket people who had wanted a complete configuration in order to determine the structural bending modes of the Atlas. However, Max Faget was strongly against installing an escape tower, deeming it unnecessary, and he won out. In the end, the Mercury-Atlas launch turned out to be a disaster, as recorded by Luge Luetjen.

"With the hangar tests completed and the flight instrumentation, parachutes, and pyrotechnics installed, Spacecraft No. 4 (MA-1) was moved to the Atlas complex on July 24, the same day that Spacecraft No. 2 arrived at [Hangar S]. Rainy weather made it difficult to complete preflight checks at the pad and caused delays and much consternation for the NASA officials there for the launch.

"The day before the scheduled date for the launch, a group of us, several in rain gear, visited the Atlas pad. The next day, early on the morning of July 29, heavy rain

Boilerplate capsule SC-5 floats in the Gulf of Mexico off the U.S. Navy School of Aviation Medicine, Pensacola, Florida. Assisted by Navy frogmen, Gus Grissom is photographed practicing egress techniques through the top of the mockup capsule. (Photo: NASA)

enveloped the Cape but the cloud ceiling soon rose high enough to be considered acceptable for launch. [John] Yardley and I were invited to be blockhouse observers. Finally, at 9:13 a.m., [NASA Operations Director] Walt Williams gave the OK to launch and the Atlas rose slowly from the launch pad. It pierced the cloud cover in seconds and the initial phases of the launch appeared normal. Then everything went wrong. Speculation had it that the Atlas either exploded or suffered a catastrophic structural failure. Whichever it was, it occurred at about 32,000 feet and a velocity of about 1,400 feet per second. It was indeed a sad day for Mercury!"[32]

After a thorough examination of all the available evidence and telemetry it was concluded that the Atlas booster had failed in the thin-skin area below the adaptor which joined the spacecraft to the booster. As a result, a stainless steel reinforcing "bellyband" was developed that wrapped around the boosters until later Atlas rockets could be manufactured with a thicker skin incorporated into this critical area.

Completed Mercury spacecraft with protective covering at McDonnell's St. Louis plant. (Photo: McDonnell Aircraft Corporation)

MERCURY SPACECRAFT NO. 2

Meanwhile, following completion of the Capsule Systems Test (CST) at St. Louis, Spacecraft No. 2 was shipped to Huntsville, Alabama where it was test-mated with the booster allocated to the MR-1 mission to ensure complete compatibility between the two. After these checks the spacecraft was airlifted to the Cape, being delivered to Hangar S on 24 July. Here it would be installed within a room-sized, protective air-filtered plastic tent. This temporary facility was nowhere near the standard of the McDonnell 'clean room,' but it went a long way towards keeping the dust and other unwanted elements at bay.

As McDonnell design engineer Jerry Roberts explains, there was essentially unlimited access to the spacecraft at this stage, which allowed a little surreptitious activity for those seeking space-flown souvenirs.

"We had access to the spacecraft all the time in Hangar S. On nights when we worked the night shift we would take advantage of this access to curl up and fold dollar bills that everyone including some astronauts had signed into the spacecraft's

With the seven Mercury astronauts watching from a bunker, an Atlas D launched into rain-soaked skies carrying the first production model of the Mercury capsule for the planned suborbital MA-1 flight. However, the Atlas exploded and disintegrated 58 seconds after liftoff. The jettisoned capsule hit the sea and was recovered, albeit extensively damaged. (Photo: NASA)

cabling, and then we'd lace the cabling back up and tape it all up out of sight. It was a pretty big deal to put something in the spacecraft and have it flown in space. The idea was for the bills to fly in space and then we would recover them when we checked out the spacecraft after the flight. Sometimes this happened, but other times we didn't get access to the spacecraft after the flight, and so to my knowledge those bills are probably still in those spacecraft wherever they are located today."[33]

Early September 1960 would prove to be a time for greater optimism in the Mercury program. The testing of Spacecraft No. 2 in the Cape's Hangar S was progressing well, and Spacecraft No. 6 had been delivered. It would be flown in February the following year on the first "bellyband" Atlas (67D) flight, as another unmanned test launch and recovery operation designated MA-2.

The badly damaged MA-1 capsule. (Photo: NASA)

The Hangar S work schedule included the installation of parachutes and pyrotechnics, following which the fully equipped Spacecraft No. 2 was transported to the pad on 26 September. The Redstone booster had arrived earlier, been erected, and was now enclosed in the gantry (or service structure) which – as for all of the Redstone launch complexes – was basically a converted oil well rig. From the time of mating with the booster through to the scheduled launch date of 7 November, everything seemed to progress smoothly and few problems were encountered.

That month, the general training of the seven astronauts also narrowed and became far more concentrated on flying the first Mercury-Redstone missions.

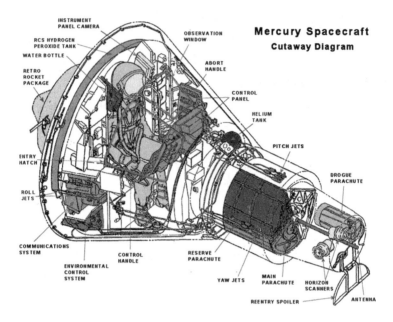

A cutaway diagram of the Mercury spacecraft. (Photo: NASA)

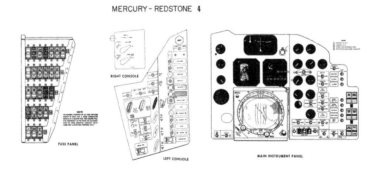

The MR-4 instrument panel. (Photo: NASA)

The MR-1 launch did not proceed on 7 November as planned, because the helium pressure in the spacecraft's control system dropped below the acceptable level. "A leak in the system, unfortunately under the heat shield, was obvious, and as a result the launch was scrubbed," Luetjen explained. "The spacecraft was removed from the booster and the heat shield dropped to expose the culprit, a leaky relief valve."[34]

The faulty valve was replaced, along with a hydrogen peroxide tank, and a minor wiring change was also made in response to an earlier test at Wallops Island. The following day, as the spacecraft was undergoing repairs, John Fitzgerald Kennedy was elected as the 35th President of the United States.

The MR-1 launch was rescheduled for 21 November and the countdown went well, apart from a short hold in order to fix a small leak in the hydrogen peroxide system. Ignition occurred at 9:00 a.m. and a mighty roar ripped across the Cape – but only momentarily. It was replaced by a sudden and unexpected silence. From behind his console in the blockhouse, Luetjen could only wonder what had gone wrong.

"Watching from the windows of the blockhouse, John Glenn and the Mercury dignitaries saw the booster wobble slightly on its pedestal and settle back on its fins after an inch or so rise. The booster engine shut down and the escape tower zipped up nearly a mile high and landed some 400 yards from the launch site. Three seconds after the escape rocket blew, the drogue package shot upward, followed in succession by the main and reserve parachutes, all of which fluttered down alongside the booster.

"After John Glenn witnessed the tower take off, he came running back to my console and said, 'My God, Luge, the tower went!' I had no appropriate answer, nor was John really expecting one. He was simply frustrated, as we all were."[35]

Two days later Robert Gilruth issued a memorandum addressed to all Mercury personnel.

"Today I received the following TWX [teletype writer exchange] from the NASA Administrator: 'As disappointed as I am in the results of yesterday's shot, I know how discouraging these troubles are to you and your fine staff. Please try to close your ears to the press comments and know that there is no lack of faith in your ability to succeed in this effort. Now is the time for real driving leadership so grit your teeth and dig in. We are solidly behind you and your outfit. Signed, T. Keith Glennan, Administrator.'

"I should like to express to the NASA and MAC staff my wholehearted agreement with the above sentiment, and my pleasure at the very fine and unstinting effort I have observed in the work here. I have every confidence that the program is sound. The recent occurrence in the MR launch attempt merely emphasizes the importance of the early flight test program in uncovering these problems which can be identified only by bringing together all the various elements of the flight system in a real exercise. [Signed] Robert R. Gilruth, Director of Project Mercury."[36]

The rocket was defueled and the remaining pyrotechnics carefully disarmed, but the fins had been damaged during the launch fiasco and the entire Redstone had to be replaced by another. Engineers tracked down the cause of the problem; they found a 'sneak circuit' in the booster ground cabling that caused an erroneous cutoff signal. Fortunately Spacecraft No. 2 was found to have come through the incident relatively undamaged and could be easily recycled. Within a week, plans were well under way for a replacement MR-1A mission using a substituted Redstone booster (MRLV-3), originally slated for the MR-3 mission. The spacecraft was fitted with the escape tower from Spacecraft No. 8 and the antenna fairing from Spacecraft No. 10.

Once the MR-1A spacecraft had been worked over at Hangar S and three verification tests completed, it was mated with the Redstone booster at Pad 5 on 9 December, with the launch set for ten days later. On the morning of 19 December, with all seven Mercury astronauts anxiously looking on, there was a 40-minute delay in the countdown caused by strong winds. And then a hydrogen peroxide solenoid valve had to be replaced, necessitating a 1-hour recycle of the countdown. Finally, at 11:15 a.m., lift-off occurred.

"This time there were no glitches," Luetjen recalled. The 83-foot Mercury-Redstone assembly was cheered on … as it lifted off and burned brightly for 143 seconds before normal cutoff."[37]

The mission was totally successful, with the Mercury spacecraft reaching an altitude of 130 miles and a range of 235 miles. The Redstone reached a slightly higher velocity than expected of 4,909 miles per hour, but this had no great impact on the overall mission. Spacecraft No. 2 was recovered from the Atlantic Ocean by recovery helicopters. "The spacecraft performed perfectly and the mission was a complete success," Luetjen said in summing up the flight. "Exuberance reigned supreme!"[38]

The MR-1A test flight had now verified the operation of the Mercury system in the space environment. At a news conference held early in 1961, Robert Gilruth praised the efforts that had gone into the creation of the Mercury spacecraft.

"In October of 1958 the Mercury vehicle was only a concept," he reported. "In two years this concept has been translated into facilities, trained teams, and flight hardware, and it is now, in two-plus years, in the initial phases of production test flights.

"This was an unusual and complex task. It required an integration of missile technology with the manned flight requirements. It involved an unprecedented cooperative effort between the military and civilians, and with foreign countries.

"It involved the building of a new technical know-how; that is, manned vehicle design and flight test methods, aeronautical unknowns, worldwide tracking and communications, and the development of industrial production and operational capability."[39]

Most importantly of all, it had involved the tremendous work and dedication of Robert Gilruth.

A CHIMPANZEE SOARS

The Mercury-Redstone 2 (MR-2) ballistic flight profile called for the Mercury spacecraft to be boosted to an altitude of approximately 115 statute miles, achieve a period of weightlessness of around 4.5 minutes and a range of approximately 290 statute miles. This time, however, there would be a living creature on board – a freckle-faced, 37-pound chimpanzee that came to be universally known as Ham.

The MR-2 flight was intended to be the final test mission of the Mercury-Redstone launch vehicle, during which a live subject closely related to humans would prove that an astronaut could function without undue difficulty in a weightlessness environment. If successful, the next mission was planned to carry one of the Mercury astronauts. Ham (an acronym drawn from Holloman Aero Medical Research Laboratory, where he got his flight training) was a chimpanzee of normally good nature and alertness, and would be the heaviest animal to be sent into space to that time, either by the Americans or the Soviet Union.

Spacecraft No. 5 had been selected for the mission, and it was fitted with six new systems that had not featured on previous flights: an environmental control system, an attitude stabilization control system, live retrorockets, a voice communications system, a "closed loop" abort sensing system, and a pneumatic landing bag.

A landing bag being installed on a Mercury capsule. Note the metallic straps attached to the heat shield. (Photo: McDonnell Aircraft Corporation)

With Ham securely sealed within a special life-support container inside the capsule, Spacecraft No. 5 lifted off from Launch Pad 5 on 31 January 1961, and landed in the Atlantic Ocean 16 minutes 39 seconds later. While essentially successful, several problems had plagued the flight. The Redstone booster over-accelerated, resulting in an earlier than expected depletion of liquid oxygen. This initiated a signal that caused the escape tower to pull the capsule free just a few seconds before it would have released normally. The overall result was that the spacecraft flew higher and over a longer range than anticipated.

Despite these problems, Ham stoically completed his assigned tasks. The spacecraft splashed down near the Bahamas, landing out of sight of the waiting recovery forces. Some 12 minutes later the first automatically generated signals were received,

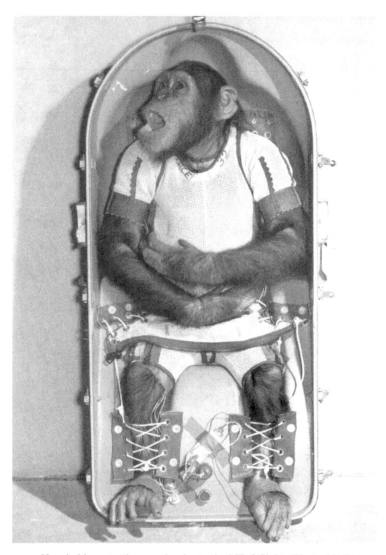

Chimpanzee Ham in his protective couch prior to the MR-2 flight. (Photo: NASA)

establishing that the spacecraft was about 60 miles from the nearest recovery ship. A search plane sighted the capsule soon after, floating upright in the ocean. When helicopters arrived they reported that the capsule was now tilted on its side, partly submerged, and seemed to be taking on water. It was later found that on impact with the water the lowered beryllium heat shield had bounced up against the bottom of the capsule, punching two holes in the titanium pressure bulkhead before tearing free. An open snorkel valve was also allowing sea water to enter the capsule. Although Ham was secured within his airtight container, it might easily have ended up as his coffin at the bottom of the Atlantic.

A helicopter crew finally latched onto the spacecraft and delivered it to the deck of USS *Donner* (LSD-20). When the spacecraft was opened, Ham appeared to be in good condition despite a little bewilderment at the attention and some wobbliness in his legs. Apart from that he was quite okay; he passed an onboard examination with flying colors and his appetite was certainly unaffected when he hungrily devoured some fresh fruit.

John King of the space agency's Public Information Office at the Cape summed up Ham's 420-mile ride through space. "Ham was going along on a pretty hectic trip, when, just one second before the end of thrust, the Redstone was burning too fast so the automatic abort system functioned. Ham got an immediate jolt of about 17G. We have a movie of it all, and he wasn't too happy when this occurred, but the amazing thing was that he griped a little, then went right back to work pushing his little levers. It was a tremendous break for the medical people. They got excellent data from this test."[40]

With many malfunctions occurring during the flight of chimpanzee Ham, it was prudently decided that the Mercury-Redstone combination was still not ready for the human passenger planned for MR-3. Despite the protests of many, the first human-tended flight was postponed pending a final Mercury-Redstone booster development flight, designated MR-BD.

Although the Soviet Union's space program was a largely unknown factor back then, in all likelihood their space chiefs would have attempted to launch a cosmonaut ahead of the United States, regardless of which Redstone flight was to carry the first American astronaut. Barring a pad explosion, the Soviets would have raced to beat an advertised launch date for the MR-3 mission, even if it meant launching the day before. They took risks, but their rocket was much more reliable as a launcher than its American counterparts. In retrospect, the decision to insert the MR-BD flight into the schedule could have ultimately cost America the prized goal of sending the first man into space, and caused a lot of consternation for the astronauts and other program participants.

Many felt that Wernher von Braun had been a little overcautious in ordering a primate test flight, and then an additional unmanned suborbital test flight prior to committing America to a manned launch. The MR-BD mission introduced after Ham's troubled ride helped to push the first flight of an astronaut into May – and allowed Soviet space scientists a little extra time to prepare for their own surprise space spectacular.

As a consequence of the problems encountered on the MR-2 mission, steps were taken to correct the problems on the now-added MR-BD flight, which would test the effectiveness of all the modifications that resulted. Carrying a boilerplate Mercury capsule loaded with a mannequin astronaut and an inert escape system, Redstone MRLV-5 roared off Launch Pad 5 at 12:30 a.m. on 24 March. With the spacecraft remaining attached as planned to the Redstone throughout the suborbital flight, the entire assembly followed the desired trajectory, achieving a peak altitude of 113.5 statute miles and a downrange distance of 307 statute miles.

This time there were no plans to recover the booster or spacecraft, which, still attached, plunged into the Atlantic Ocean at the end of a near-perfect mission. The whole operation had gone so well that there was no longer any reason to delay the much-anticipated MR-3 mission early in May.

John Glenn and Gus Grissom pose with a boilerplate replica of a Mercury capsule. (Photo: NASA)

Everything was now set for an as-yet-unnamed American astronaut to become the first person to fly into space.

However, on 12 April 1961 it was a Russian cosmonaut named Yuri Gagarin who grabbed that honor and glory with a single-orbit flight around the Earth. NASA, and the American people, were stunned and mortified. Perhaps none more so than Navy Cdr. Alan Shepard, who had earlier been secretly chosen to make the first flight of a human being into space.

In July 1961 all seven astronauts visited the McDonnell plant in St. Louis, Missouri. Back row, from left: Gus Grissom, James S. McDonnell, Alan Shepard and Scott Carpenter. Front row: Walter Burke (Vice President, McDonnell Aircraft), Gordon Cooper, Deke Slayton, Wally Schirra and John Glenn. (Photo: McDonnell Aircraft Corporation)

A pensive Wernher von Braun in his Huntsville office (Photo: NASA)

References

1. Luetjen, H.H., *Before Mercury Rose: The Half-Life of an Ex-Spaceman*, Inter-State Publishing, Sedalia, Missouri, 2001. Extracts with author permission.
2. Brownlee, Henry T. Jr., article, "Project Mercury: First Step on the Way to the Moon," *Boeing Frontiers* magazine, issue July 2009
3. Luetjen, H.H., *Before Mercury Rose: The Half-Life of an Ex-Spaceman*, Inter-State Publishing, Sedalia, Missouri, 2001
4. *Ibid*
5. Nova Online article, *To the Moon: Max Faget*, online at: http://www.pbs.org/wgbh/nova/tothemoon/faget.html
6. Cortright, Edgar M., *Apollo Expeditions to the Moon*, NASA SP-350, NASA, Washington, D.C., 1975
7. Faget, Max, interview with Jim Slade for JSC Oral History program, Houston, Texas, 18 & 19 June 1997
8. Luetjen, H.H., *Before Mercury Rose: The Half-Life of an Ex-Spaceman*, Inter-State Publishing, Sedalia, Missouri, 2001
9. *Ibid*
10. *Ibid*
11. Faget, Max, interview with Jim Slade for JSC Oral History program, Houston, Texas, 18 & 19 June 1997
12. Gilruth, Robert R., paper *Memoir: From Wallops Island to Mercury 1945–1958*, presented at the Sixth International History of Astronautics Symposium, Vienna, Australia, 13 October 1972
13. Luetjen, H.H., *Before Mercury Rose: The Half-Life of an Ex-Spaceman*, Inter-State Publishing, Sedalia, Missouri, 2001
14. Gilruth, Robert R., paper *Memoir: From Wallops Island to Mercury 1945–1958*, presented at the Sixth International History of Astronautics Symposium, Vienna, Australia, 13 October 1972
15. Swenson, Loyd S. Jr., James M. Grimwood and Charles C. Alexander, *This New Ocean: A History of Project Mercury*, NASA SP-4201, NASA, Washington, D.C., 1989
16. Roberts, Jerry, private memoir CD, *Early Space Program Memories*, recorded from 3 June 2012. Extracts taken with permission
17. *Toledo Blade* (Ohio) newspaper, unaccredited article, "Brush Beryllium to Make Man's Space Capsule Shield", issue 8 June 1959, pg. 11
18. *Victoria Advocate* (Texas) newspaper, unaccredited article, "Project Mercury Shield Forged by Alcoa, Brush," issue 16 July 1959, pg. 24
19. Carpenter, Malcolm S., Cooper, L. Gordon, Jr., Glenn, John H. Jr., Grissom, Virgil I., Schirra, Walter M., Jr., Shepard, Alan B. Jr., and Slayton, Donald K., *We Seven*, Simon and Schuster Inc., New York, NY, 1962
20. Thomas, Shirley, *Men of Space* (Vol. 3), chapter "Alan B. Shepard," Chilton Company, Philadelphia, PA, 1961, Pg. 197
21. *The Age* (Melbourne, Australia) newspaper, unaccredited article, "A Fleet of Manned Space Ships in the Making", issue Tuesday, 26 April 1960

22. *Ibid*
23. Faget, Max, interview with Jim Slade for JSC Oral History program, Houston, Texas, 18 & 19 June 1997
24. *Toledo Blade* (Ohio) newspaper, Elsie Cram article, "Astronauts Burn Midnight Oil," issue 1 March 1960, pg. 4
25. Luetjen, H.H., *Before Mercury Rose: The Half-Life of an Ex-Spaceman*, Inter-State Publishing, Sedalia, Missouri, 2001
26. Faget, Max, interview with Jim Slade for JSC Oral History program, Houston, Texas, 18 & 19 June 1997
27. NASA Specification No. S-6 "Specifications for Manned Space Capsule, revised 26 January 1959, Section 3.5.3 "Entrance and Emergency Egress Hatch."
28. Luetjen, H.H., *Before Mercury Rose: The Half-Life of an Ex-Spaceman*, Inter-State Publishing, Sedalia, Missouri, 2001
29. Roberts, Jerry, private memoir CD, *Early Space Program Memories*, recorded from 3 June 2012
30. Thomas, Shirley, *Men of Space* (Volume 4), chapter "Robert R. Gilruth," Chilton Company, Philadelphia, PA, 1962, pg. 64
31. Luetjen, H.H., *Before Mercury Rose: The Half-Life of an Ex-Spaceman*, Inter-State Publishing, Sedalia, Missouri, 2001
32. *Ibid*
33. Roberts, Jerry, private memoir CD, *Early Space Program Memories*, recorded from 3 June 2012
34. Luetjen, H.H., *Before Mercury Rose: The Half-Life of an Ex-Spaceman*, Inter-State Publishing, Sedalia, Missouri, 2001
35. *Ibid*
36. Gilruth, Robert R., "Memorandum for All Mercury Personnel," NASA Space Task Group memo, Langley Field Virginia, 23 November 1960.
37. Luetjen, H.H., *Before Mercury Rose: The Half-Life of an Ex-Spaceman*, Inter-State Publishing, Sedalia, Missouri, 2001
38. *Ibid*
39. Thomas, Shirley, *Men of Space* (Volume 4), chapter "Robert R. Gilruth," Chilton Company, Philadelphia, PA, 1962, pg. 63
40. Thomas, Shirley, *Men of Space* (Vol. 3), chapter "Alan B. Shepard," Chilton Company, Philadelphia, PA, 1961, Pg. 202

2

An astronaut named Gus

Gus Grissom never seemed to fit the archetypal American hero mold. A stocky and somewhat stubby man who stood at 5 feet 7 inches, he looked more like the neighborhood motor mechanic or television repairman than an astronaut. But he excelled as an Air Force test pilot and as a Mercury astronaut, becoming an integral part of NASA's drive to the Moon. While he may not have been the most sociable or loquacious member of the astronaut group, he was well respected by them. "Gus was a very bright young man who didn't have a lot to say most of the time," fellow astronaut Scott Carpenter told the author in 2013, "but when he said something it was of great value and always worth listening to."[1]

A MAN OF FEW WORDS

The late John A. ('Shorty') Powers, former NASA Public Affairs Officer would have agreed with Carpenter's characterization. "Gus is the quiet one," he once observed. "He doesn't talk much, but when he does speak, the words come out in short bursts – like a fighter pilot's measured use of limited ammunition. When he fires off a burst, one had better be listening carefully, because he's only going to say it once and there won't be any surplus words."[2]

Fellow Mercury astronaut Wally Schirra had a good grasp on the personality of Gus Grissom, saying he brought a vast amount of knowledge and experience into the space program, and his opinions as an extremely capable and competent test pilot and engineer were highly valued and respected. "Gus did not consider himself as the hero type, nor was he impressed with personal prestige. He was a quiet, unassuming, and completely unpretentious person, and his reasons for wanting to participate in this venture were really quite basic. Should the officials at NASA share his belief that he was one of the better qualified people for this new mission, then he was proud and happy to help out. Although Gus was the shortest of any of us chosen in that first group of astronauts, his physical stature did not in any way hinder or inhibit his

C. Burgess, *Liberty Bell 7: The Suborbital Mercury Flight of Virgil I. Grissom*, Springer Praxis Books, DOI 10.1007/978-3-319-04391-3_2, © Springer International Publishing Switzerland 2014

enormous competitive spirit. He possessed a strong desire to succeed in everything he undertook, and this unbeatable desire to win was matched only by his determination and perseverance to see a job through to its satisfactory conclusion."[3]

The family name Grissom actually evolved from England and the surname Gresham. According to genealogists the Greshams came to America from Surrey, England, and later chose to distinguish themselves from the loyalists by changing their name to Grissom. The first Gresham to immigrate to America was John Gresham, who, with his wife and son, settled in Arundel County, Maryland in the mid-1600s. For Gus Grissom it was a similarly long and difficult trek from Mitchell, Indiana to flying into space, but his tenacity and a driving urge to go beyond any limitations imposed by others was always an integral part of his character.

Virgil Ivan Grissom was born at 8:00 a.m. on 3 April 1926 in the small mid-western city of Mitchell in southern Indiana, the second child of Dennis and Cecile King Grissom. In a significant and somewhat connective sense, that same day American rocket scientist Robert H. Goddard conducted a second successful launch of a liquid-fueled rocket at his Aunt Effie's farm in Auburn, Massachusetts.

Grissom's father was a signalman for the Baltimore and Ohio Railroad, while his mother was a homemaker. An older sister had died in infancy before his birth, and he was followed in turn by three younger siblings, Wilma, Norman and Lowell. The family lived in a simple, white-frame house at 715 Baker Street (later to be renamed Grissom Street). He took his early education at Mitchell's Riley Elementary School, a short walk from his house, and while he possessed an IQ said to be around 145 he was only an average student and had no real plans for the future. He did, however, become

The Grissom home, circa 1968 (Photo: Carl L. Chappell)

moderately interested in flying airplanes. "I guess it was a case of drifting and not knowing what I wanted to make of myself," he said. "I suppose I built my share of model aeroplanes, but I can't remember that I was a flying fanatic."[4] As a child he attended the local Church of Christ where he remained a lifelong member and later joined Beaver Patrol with the local Boy Scout Troop 46, developing his enduring love of the outdoors.

Every morning, in order to have a little pocket money for his own activities, young Gus would make his way to the downtown bus station and collect that day's edition of the *Indianapolis Star* newspaper for his delivery route. In the evenings he would also pick up and deliver the local newspaper, the *Bedford Times*.

In 1940 Grissom was enrolled at Mitchell High School, where he soon found to his chagrin that his short stature precluded him from playing varsity sports. Instead he became a fierce competitor in the school's swimming pool. While he could not play basketball for his school, he took immense pride in being a member of the Boy Scout Honor Guard, which presented the American flag before any games. While engaged in this activity during one game, he caught the eye of fellow student Betty Lavonne Moore, who played the drum in the school band. When he came and sat with her during the half-time break, Betty realized to her delight that the attraction was mutual. "I met Betty Moore when she entered Mitchell High School as a freshman," Grissom later admitted, "and that was it – period, exclamation point!"[5]

JOINING THE AIR FORCE

While at Mitchell High School, Grissom completed a year of pre-cadet training in the U.S. Army Air Corps, which he found most enjoyable. By this time his interest in aviation had taken a deeper hold, and he took on summer casual work in order to pay for brief flights in barnstorming airplanes at nearby Bedford airport, Indiana. A local attorney who owned a small aircraft would often take him on flights for a one dollar fee and taught him the basics of flying.

Grissom picked up the nickname 'Gus' during a card game when someone saw the abbreviated name on an upside-down score card and mistakenly translated it to "Gus." Before long, Grissom's friends also began calling him Gus, and it stuck. But he will always be known as Virgil to the people in his hometown in Mitchell.

World War II broke out while Grissom was still in high school, and he was eager to enlist upon graduation. On 8 August 1944 – Betty's seventeenth birthday – he was inducted into the Army Air Forces at Fort Benjamin Harrison, with the expressed desire of becoming a pilot. He was subsequently ordered to Sheppard Air Force Base (AFB) in Wichita Falls, Texas, for five weeks of basic training. Then he was assigned to Brooks Field in San Antonio, where, to his extreme disappointment, he spent his days behind a desk as a lowly clerk.

Grissom took some short leave and on 6 July 1945, while still in his teens, he and Betty were married in the First Baptist Church in Mitchell. He then returned to the Air Force while Betty remained in Mitchell, working at the Reliance Manufacturing Company making shirts for the Navy. Soon after, Japan capitulated and the Second

Newlyweds Gus and Betty Grissom. (Photo: World Book Science Service)

World War came to an end. Dispirited with the lack of flight training within the Air Force, Grissom left the service in November 1945 with the rank of corporal and took up a job fitting out school buses in Mitchell's Carpenter Body Works, but it was the kind of mundane and repetitive work he hated. Deciding to become a mechanical engineering student, he enrolled at Purdue University, Indiana under the G.I. Bill in September 1946. He and Betty took a small apartment near the campus, and while Betty worked as a long-distance telephone operator to help pay the bills, he found some after-class work as a short-order cook "frying hamburgers for 30 hours a week."[6] Fortunately he found his studies absorbing and to his liking, and he graduated with his bachelor of science degree in February 1950.

He had contemplated entering private industry at this stage of his life, but when the Korean war broke out Grissom decided to re-enlist in the Air Force and was assigned to Randolph AFB, Texas as an aviation cadet. On 16 May 1950, he and Betty welcomed their son Scott into the world. In September Grissom graduated from basic flight training and was sent to Williams AFB in Phoenix, Arizona for more advanced training. He received his wings and was commissioned a 2nd lieutenant in March 1951. In December of that year he was shipped off to the conflict in Korea to fly North American F-86 Sabre jets with the 334th Fighter Interceptor Squadron.

2nd Lieutenant Grissom after receiving his wings in March 1951. (Photo: Carl L. Chappell)

Six months after his arrival in South Korea Grissom had reached the 100-mission mark and was promoted to 1st lieutenant. He was eager to fly another 25 missions, but his request for an extension was refused and he returned home having earned a Distinguished Flying Cross and the Air Medal with cluster. After a period in Flight Instructor School he was designated as a flight instructor at Bryan AFB, Texas. On 30 December 1954 he and Betty completed their family with the birth of second son, Mark. The following year Grissom was assigned a place at the Air Force Institute of Technology at Wright-Patterson AFB, Ohio, to study aeronautical engineering. He then won an assignment to the prestigious and highly prized Test Pilot School at Edwards AFB, California, checking out advanced-design fighter airplanes.

THE MAKING OF AN ASTRONAUT

Following his graduation from Test Pilot School, Grissom, now bearing the rank of captain, returned to Wright-Patterson AFB in May 1957 as a test pilot assigned to the fighter branch.

An exultant Grissom after completing his 100th combat mission during the Korean war. (Photo: World Book Science Service)

One day in 1958, an adjutant handed Capt. Grissom an official teletype message marked "Top Secret," instructing him to report to an address in Washington, D.C., and to wear civilian clothing. There were no other details, but he knew there was a challenge in there somewhere. As it turned out, he was one of 110 carefully selected candidates who had met the general qualifications for astronaut training. They would undergo initial briefings and medical screening in the quest to find America's first astronauts for NASA.

After attending the briefing, in which the attendees were given information about Project Mercury, they were offered a crucial choice. If they decided to volunteer for the chance to become what NASA referred to as an "astronaut", they would move onto the next phase of the selection process. If not, then they could return without prejudice to their present service. Some turned down the chance to be involved

Gus Grissom at the U.S. Air Force Test Pilot School, California. (Photo: U.S. Air Force)

in this new venture. There were too many unknowns and they preferred to continue with the work they were already involved in. Grissom now had to think seriously about his own future.

"It was a big decision for me to make. I figured that I had one of the best jobs in the Air Force, and I was working with fine people. I was stationed at the flight test center at Wright-Patterson, and I was flying a wide range of airplanes and giving them a lot of different tests. It was a job that I thoroughly enjoyed. A lot of people, including me, thought the [Mercury] project sounded a little too much like a stunt than a serious research program. It looked, from a distance, as if the man they were searching for was only going to be a passenger. I didn't want to be just that. I liked flying too much. The more I learned about Project Mercury, however, the more I felt I might be able to help and I figured that I had enough flying experience to handle myself on any kind of shoot-the-chute they wanted to put me on. In fact, I knew darn well I could."[7]

Afterwards, when he told Betty about Project Mercury and the chance that was being presented to him, she said that he would have her full support in whatever he decided to do. After a lot of thought, Grissom decided to volunteer, following which he was subjected to intense physical and psychological testing through early 1959. At one stage he came close to being disqualified when doctors discovered that he suffered from hay fever, and he had to convince them that it would not bother him in space. He argued that he would be sealed in a pressurized spacecraft, with no pollen present. It must have been a close call, as there was a tremendous emphasis on physical fitness. With his usual determination he won his case.

On Thursday evening, 2 April 1959, Gus Grissom received the phone call that would change his life forever. On the other end of the line was NASA's assistant manager for the project, Charles Donlan, who officially informed him that he had been selected as one of the space agency's seven Mercury astronauts.

"After I had made the grade, I would lie in bed once in a while at night and think of the capsule and the booster and ask myself, 'Now what in hell do you want to get up on that thing for?' I wondered about this especially when I thought about Betty and the two boys. But I knew the answer: We all like to be respected in our fields. I happened to be a career officer in the military – and, I think, a deeply patriotic one. If my country decided that I was one of the better qualified people for this new mission, then I was proud and happy to help out. I guess there was also a spirit of pioneering and adventure involved in the decision. As I told a friend of mine once who asked me why I joined Mercury, I think if I had been alive 150 years ago I might have wanted to go out and help open up the West."[8]

Following the announcement of the names of the seven Mercury astronauts in Washington on 9 April 1959, they became instant celebrities – something that caught them (and NASA) completely unawares. "It happened without us doing a damn thing," Deke Slayton later mused. "We show up for a news conference … and now we're the bravest men in the country. Talk about crazy!"[9]

SOME PERSONAL REFLECTIONS

An air of mystique quickly surrounded the Mercury astronauts. People eagerly sought out and consumed information about them and their families in newspapers, especially through *Life* magazine, which was awarded sole rights to their stories under a mutually beneficial deal thrashed out between the astronauts' tax attorney Leo D'Orsey and the magazine.

Perhaps the most enigmatic of the seven was Gus Grissom, who tried to avoid the press and public speaking whenever he could, although he abided this as part of his duties as a NASA astronaut. He was more at home in the cockpit of a fighter jet or poking around inside a space capsule than he was in revealing details of his private life. A man of few words, he was quickly given the rather unfair sobriquet of "Gruff Gus." He didn't care; he had a job to do, and he disliked distractions.

One of the lesser-liked duties for most of the Mercury astronauts – particularly Grissom – involved giving speeches. But on one occasion he unwittingly created a humorous, oft-repeated story that would always follow him around. It came about when all seven astronauts were on a tour of the Convair plant in San Diego to see Atlas rockets under construction. At one stage of their visit the astronauts were seated on a podium outside the plant, in front of 18,000 cheering employees. A Convair executive made a short speech, then turned and asked if an astronaut would care to respond. Grissom somehow found himself propelled forward to the microphone by his fellow astronauts, but he had been caught off-guard and was unprepared. As the tumultuous applause died down he cleared his throat and suddenly developed a dose of stage fright. He hesitated, raised his hands for quiet and blurted out, "Well…do good work!" He then rejoined the other grinning astronauts. Momentarily, the crowd was silent, and then they burst into wild clapping and cheering. It was as if he'd just recited the Gettysburg address, and the workers loved him. "Do good work!" – it now became their mantra and mission statement. Gus's stammering slogan was stitched onto a huge banner, which was hung above the plant's work bay to serve as an inspiration to all.

Another side of Gus Grissom was recalled by Frank Moncrief from Virginia Beach, Virginia, who shared a particularly vivid memory of the astronaut during a training session.

A poster inspired by Gus Grissom's three words to the Convair employees. (Photo: NASA)

"The original seven astronauts were sent to Little Creek Naval Amphibious Base so the Navy frogmen could give them scuba lessons. It was believed that weightlessness in water was similar to weightlessness in space. I was honored to be one of the frogmen picked to be an instructor.

"These guys were the sharpest of the sharp. John Glenn asked me how the Aqua-Lung valve worked, and I explained in detail the valve's functions. That did not satisfy him. I had to take the valve apart, explaining every screw, every spring and every diaphragm. They were that thorough. For graduation, the astronauts entered the pool with a full tank of air and were told to swim around without coming up for anything. We instructors began harassing them, pulling off a face mask or fins, pulling out mouthpieces, turning off their air. Then we brought them to the surface and congratulated them. I was proud to have been a part of that. But it wasn't over.

"A couple of weeks later, we got word that the astronauts were inviting the instructors to ride in a jet. When we arrived at Langley, the astronauts greeted us like old buddies. I was given Deke Slayton's helmet and Walter Schirra's flight suit to wear. My pilot was Gus Grissom, and we boarded a T-33 jet trainer. He told me to put my left hand on my left knee and my right hand on my right knee and not to touch anything. I did as I was told.

"Gus started to describe everything: 'I am going to start my rollout; when the aircraft gets to 180 knots I will lift this tub off the ground.' (These guys were used to flying high-performance aircraft, so flying a T-33 was demeaning.) He threw the plane into a steep left-hand bank – WOW! The g's! My helmet felt like it weighed a ton. Then Gus straightened out the plane and said there was a hole in the clouds. WOW, again! We went from 3,000 to 10,000 feet in seconds. The clouds looked like lava-lamp spirals. Gus began to fly around these spirals, up and down. He asked me how I was doing, then said, 'Frank, don't you dare throw up in my plane.' I answered, 'Gus, just before I die, I am going to throw up in this plane.' (I was close.) Then we leveled off, to my relief.

"Gus said, 'There's Richmond down there.' I said, 'How can you tell from up here?' He said, 'Let's go take a look.' We went straight down. WOW – more g's and g's. Then, thank God, he leveled off again.

"When we landed back at Langley, we got out of the plane, and Gus put an arm across my shoulders, pointed up and said, 'Frank, that's *my* pool.'

"I answered: 'Touché, Gus.'"[10]

All of the Mercury astronauts were highly competitive individuals, even amongst themselves. The late Cece Bibby was an exceptionally gifted artist who found fame as the person who painted insignias on the sides of the orbital Mercury spacecraft, and she recalled that "each of them wanted to have the first flight. It didn't matter that the first couple of flights would be suborbital; first was first and that was part of the attraction.

"These guys had really fought to be named the first astronauts, although some people referred to them as astronaut candidates. Wally Schirra once said that they were actually only 'half-astronauts' until a space flight was made. So, after the first seven were picked they then fought to make the first flight. This led to a lot of good-natured competition and jockeying for position and it involved every aspect of their flights.

"When Al made his flight there was a stencil cut for the name *Freedom 7* and the name was sprayed onto the capsule. The same was true for Gus's *Liberty Bell 7*. I don't know who sprayed the names on the capsules. I do know that when John Glenn decided he wanted his *Friendship 7* hand painted on his capsule there was a good bit of 'joshing' that went on about it. Al and Gus made comments that a stencil wasn't good enough for John; that he had to have his name hand painted by an artist.

"Gus told me later that he wished he'd have had an artist do his *Liberty Bell*. He said it really bugged him that someone else thought of it and he hadn't. Competitiveness."[11]

That same competitiveness also involved fast cars and the pulling of pranks – or "gotchas" – on each other. Grissom's brother Lowell recalled one hilarious incident.

"It was late evening, and [as] Gus exited the Cape's gates, he drove his Corvette at its usual speed of 100 miles per hour. He got on the first highway, and one cop picked him up and started chasing him. A little further away, a second police car joined the chase. And by the time Gus reached Cocoa Beach, three police cars were following him.

"Gus was far ahead of them, and when he reached the motel where he was staying he parked his Corvette in front of Alan Shepard's room, then walked into his own. The cops came and saw the Corvette sitting there and felt the hood and said, 'Yeah, this is it.' And they banged on Shepard's door. Shepard comes to the door, half-asleep, and they pull him out, throw him on the ground and cuff him. Meanwhile, Gus had changed into pajamas and was watching from his room.

Cece Bibby painting the Friendship 7 logo onto the side of John Glenn's spacecraft. (Photo: NASA)

"As the police officers handcuffed Shepard, Gus yelled at them out the window. 'Hey, guys, can you hold it down out there? Some of us have to go to work in the morning!'"[12]

Whatever his faults, or the perception others might have of him, no one could deny the application, thoroughness, and expertise that Gus Grissom brought to his work and training. He gained everyone's respect, and two years after the selection of the Mercury astronauts his dedication to the task was rewarded by the assignment to his first flight into space.

References

1. Scott Carpenter interview with Colin Burgess, Spacefest V, Tucson, AZ, 4 May 2013
2. Thomas, Shirley, *Men of Space, Vol. 7*, chapter "Virgil I. Grissom," Chilton Books, Philadelphia, PA, 1965, p. 97
3. Chappell, Carl L., *Seven Minus One: The Story of Astronaut Gus Grissom*, New Frontier Publication, Mitchell, IN, 1968, p. xi
4. Indiana Historical Society, Virgil "Gus" Grissom, (undated) by IHS staff, online at *www.indianahistory.org/our-collections/library-and-archives/notable-hoosiers/virgil-gus-grissom*
5. Grissom, Virgil, *Gemini: A Personal Account of Man's venture into Space*, The Macmillan Company, Toronto, Ontario, 1968, p. 19
6. *Ibid*
7. Burgess, Colin and Kate Doolan. *Fallen Astronauts: Heroes Who Died Reaching for the Moon*, University of Nebraska Press, Lincoln, NE, 2003, p. 95
8. Carpenter, Malcolm S., Cooper, L. Gordon, Jr., Glenn, John H. Jr., Grissom, Virgil I., Schirra, Walter M., Jr., Shepard, Alan B. Jr., and Slayton, Donald K., *We Seven*, Simon and Schuster Inc., New York, NY, 1962, pp. 73–74
9. Schefter, James, *The Race: The Uncensored Story of How America Beat Russia to the Moon*, Doubleday Books, New York, NY, 1999, p. 67
10. Moncrief, Frank, article "Lessons Learned" for the *Virginian-Pilot* newspaper, Norfolk, VA, interviewed by Diane Tennant, 22 September 2012
11. Bibby, Cece, *Cece Bibby's Genealogy*, online stories at *http://freepages.genelogy.rootsweb.com/~cecebibby/nasa-stories/naked-lady.htm*
12. Grissom, Lowell. Online story at *http://baileyloosemore.wordpress.com/2012/06/22/honoring-a-hero*

3

Preparing for launch

On Tuesday, 21 February 1961, NASA finally released the names of John Glenn, Gus Grissom and Alan Shepard as having been selected as the prime candidates to enter the special training required for the final stages of preparedness for the first suborbital Mercury mission.

At the same time that the three names were announced, Gus Grissom was on duty at the NASA tracking station in southernmost Bermuda, where he was sitting at a control console during the unmanned Mercury-Atlas (MA-2) suborbital space shot, the main purpose of which was a particularly rugged reentry test of the capsule's heat shield. The tracking station was on Cooper's Island, a 77-acre rock-and-coral shelf in the Atlantic, some 600 miles from the United States.

The MA-2 flight was launched that day from Complex 14 at Cape Canaveral and flew a suborbital mission lasting 17 minutes and 56 seconds. Atlas rocket 67D carried Spacecraft No. 6 to an altitude of 114 statute miles at a speed of 13,227 miles an hour. All the test objectives of the flight were achieved, and the capsule was recovered 1,432 miles downrange.

FIRST SPACEMAN

Grissom was pleased with the overall success of the MA-2 flight, but his mother was somewhat less thrilled when the news broke that he had been selected as one of three candidates to make America's first flight into space. "Oh no," she told reporters. "I've been hoping and praying he wouldn't be the one. I hate to be against him because I know he wants to go." After a pause, she added, "I've thought all along he would be one of them. And I'll be more than proud if everything turns out alright."[1]

For his part, Grissom had been brimful of confidence that he might get the first flight, but that lofty ambition had been abruptly thwarted during a private meeting between the astronauts and their boss, Robert Gilruth, on 19 January during which Gilruth had unceremoniously told the seven astronauts that after much deliberation

C. Burgess, *Liberty Bell 7: The Suborbital Mercury Flight of Virgil I. Grissom*, Springer Praxis Books, DOI 10.1007/978-3-319-04391-3_3, © Springer International Publishing Switzerland 2014

Shepard had been chosen to make the first suborbital flight and Grissom the second, with Glenn backing up both flights. They were asked to keep the news to themselves until the time of the actual first mission, a chore that was often difficult. As Grissom stoically told one newsman when asked if he would like to be the one chosen to make the first flight, "Everything I do is influenced by it. With everything I do, I expect it. I am here to ride the capsule."[2]

The three Mercury candidates for the first suborbital flight: Glenn, Grissom and Shepard. (Photo: NASA)

NASA's Mercury tracking station on the island of Bermuda. (Photo: NASA)

Spacecraft No. 11, which would fly the MR-4 mission, was delivered from the McDonnell plant on 7 March. Grissom had viewed the capsule under assembly at the St. Louis plant two months earlier and attended several production meetings there as the months rolled by prior to the capsule being delivered to the Cape. "I thought it would be good for the engineers and workmen who were building my spacecraft to see the pilot who would have to fly it hanging around," he observed in the astronaut book, *We Seven*. "It might make them just a little more careful than they already were and a little more eager to get the work done on time if they saw how much I cared."[3]

Following NASA's announcement, speculation on which of the three astronauts would fly the first mission was rife in the press. On 12 March that year, *Aviation Week & Space Technology* magazine editor Marvin Miles published an article under the title "Marine Stands Out as Astronaut Choice." The article began by explaining its rather emphatic title, suggesting, "We say this because Glenn, at 39 the eldest of the group, has always been the father of the seven-man team; a leader without appointment; an officer particularly respected among the astronauts – and apparently all others in the Mercury program – for his personality, his dedication, his skill and his experience." While incorrect in its speculation, the magazine was only echoing the public's prevailing conjecture on the subject.[4]

AN UNNERVING INCIDENT

On 24 March, the Redstone rocket had been successfully rated for human use with the completion of the Mercury-Redstone Booster development (MR-BD) suborbital proving flight, carrying a previously flown and unmanned, non-functional Mercury "boilerplate" capsule. However, before a Redstone booster carried an astronaut on a similar flight it was deemed necessary to conduct an orbital test of the Mercury-Atlas combination as a means of checking out the Mercury spacecraft in the actual orbital environment. This MA-3 launch would take place on 25 April. The only subject to be carried on board Mercury spacecraft No. 8 was a mechanical astronaut, known to the NASA astronauts as an "astro-robot."

For the MA-3 launch, Gus Grissom and fellow astronaut Gordon Cooper were assigned the grandstand spectacle of flying delta-wing F-106A jets over the Cape and keeping company with the Atlas rocket as it gathered speed after liftoff.

"I was to approach MA-3 at 5,000 feet, ignite my afterburner, and climb up in a spiral alongside to observe this early phase of its flight," Grissom wrote in his memoir, *Gemini*. "Gordon Cooper would take over from his 25,000 feet level and continue observation of the big bird."

Grissom then noted that everything seemed to go perfectly on launch day. "To allow me more observation time, it was decided I should go in at about 1,000 feet, keeping about half a mile distant from MA-3 after liftoff. No sweat."

A successful liftoff for the MA-3 mission. (Photo: NASA)

Ignition and liftoff occurred right on schedule, and Grissom was having a dream flight climbing along with the ascending Atlas rocket. As he watched, he saw to his surprise the escape tower fire unexpectedly, prematurely hauling the Mercury capsule away from the Atlas. Things had begun to go seriously wrong when the Atlas failed to follow its pitch and roll programs. At 43 seconds into the flight the Range Safety Officer at the Cape decided to terminate the erratic path of the missile by transmitting the destruct command. This prompted the escape tower to react and haul the space-craft clear moments before the Atlas blew apart. A shocked Grissom later recorded the scene before him as "*Kablooie*! The biggest fireball I ever want to see!"

His pilot's reactions instinctively came into force, and Grissom pulled over and away from the massive explosion, but spectators watching on the ground feared his aircraft had flown straight into the conflagration. A friend later told Grissom he had turned to his shocked wife on Cocoa Beach and mournfully stated, "Well, now there are only six astronauts."

Gordon Cooper, stationed above at 15,000 feet, was horrified to see the ascending rocket erupt into a gigantic fireball right below his F-106. He later reported that the soaring escape tower and capsule missed his aircraft by what seemed to be 15 feet. Somehow, neither F-106 suffered any damage following the explosion.

Despite the scare, Grissom quickly recovered and decided to follow the released capsule as it descended on its parachute to the water. The test engineer in him had clicked in, and he knew that NASA technicians and others would welcome a report on this phase of the aborted flight. But there was another shock in store.

"I remember thinking, my gosh but these are big seagulls around here today. They were flying all around my plane. And *then* it hit me – these were no seagulls. They were chunks of the exploded Atlas, falling." His luck held, as none of the tumbling chunks of rocket hit his aircraft. "It was quite a spectacle," he noted, "but never again, thanks."[5]

TRAINING INTENSIFIES

On 4 April, Glenn, Grissom and Shepard left for Pennsylvania to undertake refresher tests at the Naval Air Development Center in Johnsville, just outside of Philadelphia. The center had once served as an aircraft production factory but was later converted to U.S. Navy research laboratories that were studying pilotless aircraft, electronics and weapons. It was also home to a vast human centrifuge building, which assisted in researching the limits of a pilot's tolerance to a rapid buildup of gravitational or g-forces.

Here the future astronauts were strapped into a 10-foot by 6-foot steel gondola situated at the end of a 50-foot arm, and secured in various positions relative to the applied g-force. As well, the gondola could be rotated by controllers while the high-performance centrifuge was in action. At the far working end of the centrifuge's arm was a 4,000 horsepower electric engine to hurl the gondola and its hapless occupant around the circular chamber at high speed.

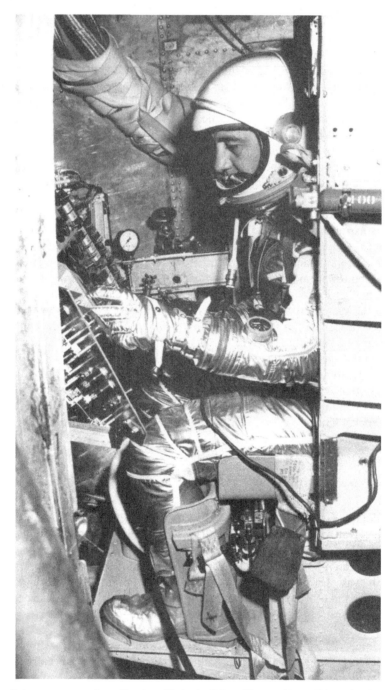

Grissom prepares for a dizzying ride in the Johnsville centrifuge. (Photo: NASA)

John Glenn at work in the flight procedures trainer. (Photo: NASA)

John Glenn would later refer to the Johnsville centrifuge as a "dreaded" and "diabolical" part of astronaut training. In his book, *John Glenn: A Memoir,* he said, "Whirling around at the end of that long arm, I was acting as a guinea pig for what a human being might encounter [whilst] being launched into space or reentering the atmosphere. You were straining every muscle of your body to the maximum … if you even thought of easing up, your vision would narrow like a set of blinders, and you'd start to black out."[6]

On that occasion Grissom made two simulated Mercury acceleration profiles, which proved to be his last preflight experience on the dreaded centrifuge.

Another particularly vital training aid frequently used by the astronauts was the flight procedures trainer. A complex device, it comprised a mockup version of the Mercury capsule with all of its systems connected to exterior control panels and computers. The trainer allowed the astronauts to test their proficiency by flying simulated missions and learning how to control possible contingencies such as emergency situations.

Another valuable training device was the Air-Lubricated Free Attitude (ALFA) trainer at NASA Langley. In using the ALFA, the astronaut first strapped himself into a couch that was then finely balanced on a cushion of compressed air in order to remove any feeling of friction. Then, moving very freely on all three axes, it provided the astronaut with important practice in lightly maintaining their spacecraft at the correct orbital attitude. "This trainer provided the only training in visual control of the spacecraft," Gus Grissom would later recall.[7]

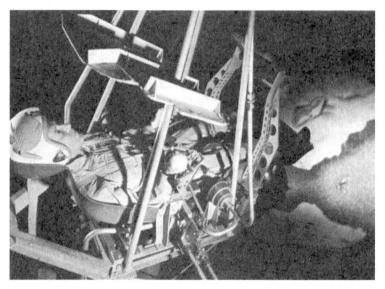

Wally Schirra takes his turn on the ALFA training device. (Photo: NASA)

FIRST TO FLY

As the first months of 1961 passed, most of Grissom's attention was centered on the forthcoming suborbital flight of Alan Shepard for which he was the backup pilot. He stood ready to take over the role of prime pilot in the event that Shepard was unable to fulfill that role. In addition to assisting Shepard throughout the preparations for the flight, Grissom worked with him through the various delays that plagued the MR-3 launch.

On 5 May 1961 it all came together for NASA when Shepard became the first American to be launched into space on a suborbital trajectory, thereby confirming – albeit briefly – that a human being could not only survive the dynamics of launch and reentry, but work without any physiological hindrance in weightlessness. His *Freedom 7* capsule came through the 15-minute flight with flying colors, and the Redstone booster performed with its renowned reliability.

"We all pulled together on that one," Grissom later wrote, "just as we do on all our flights, and I stuck pretty close to Al until he went up. I was with him when he dressed, and I rode out in the van with him to the pad. After that flight I buckled down to my own problems and stayed near my own capsule as much as I could."[8]

Approximately 60 days prior to the MR-4 mission, with the launch tentatively set for mid-July, Grissom began specific preparation for the flight, and his experiences during this lengthy buildup period would eventually shape the pattern of his actions and responses during the flight. During this period he mostly stayed at the nearby Holiday Inn in Cocoa Beach so as to closely follow the preparation of the capsule and booster at the Cape, flying home briefly each weekend to visit Betty and their two boys. He participated in countless systems checks carried out in the Hangar S test

Both in dress uniform, Grissom stands alongside America's first man in space, Alan Shepard. (Photo: NASA)

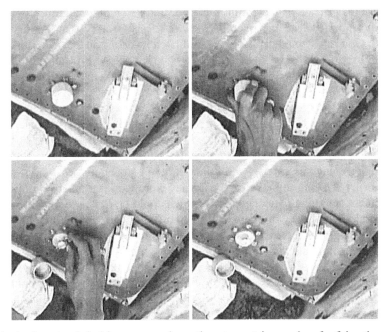

Clockwise from top-left, this sequence shows the astronauts' procedure for firing the explosive hatch. The first photo shows the knurled knob that had to be turned and removed (as seen in the second image) to reveal the plunger. And once the pin had been removed, the plunger was ready to be depressed with a solid push. (Photos: McDonnell Douglas)

facility, and most of his waking hours were spent going over every component again and again with McDonnell and NASA engineers and scientists assigned to the test.

"It's good for them to know that the guy who is going to ride it is around," Grissom commented of this period, echoing his earlier philosophy on working alongside the McDonnell people while Spacecraft No. 11 was being assembled in St. Louis.[9] He also took up jogging along the beach in order to keep himself in good physical trim and mentally run through any mission-related issues. As a precaution against sustaining any disqualifying injuries, he temporarily gave up his favorite sport of water-skiing, and moderated his driving speed to stay with lawful limits.

CHECKS AND REVIEWS

From 14-17 May, Grissom participated in Systems Engineering Department Report (SEDR) 83, a capsule pressure chamber test. This was the only chance for Grissom to familiarize himself with the operation of the environmental control system under simulated flight conditions. It also provided an opportunity for physicists to gather baseline metabolic data on him. And over the following three days, 18-20 May, he participated in SEDR 61, the communications systems check that permitted him to check out all of his communications equipment.

Next, from 1-3 June, Grissom took part in SEDR 73, the capsule's manual reaction control systems tests. This gave him the opportunity to manipulate the control stick and become familiar with the overall 'feel' of the control system. As his training intensified, Grissom also ran a total of 12 simulated missions on the procedures trainer from 17-23 June, followed by 24 simulated missions on the ALFA trainer on 28 June.[10]

While all of this advanced training was taking place at the Cape, there were also daily scheduling meetings to attend. These sessions kept everyone apprised of the flight's progress and any problems that had cropped up. As Grissom reflected in *We Seven*, this was where they reviewed any work being done on the various systems.

"It was also here that the perfectionists in the crowd would sometimes try to stop the show and redesign the whole system again from scratch. I wanted a safe and efficient capsule as much as anyone, and I did not blame the engineers, who were proud of their work, for trying to make each part they were working on absolutely perfect. I knew that if something happened to me which could be traced to one of their decisions, it would hang heavy on them. But I had also noticed, during my days as a test pilot, that engineers are seldom satisfied to have something work *well*. They often want to go on testing and testing a system until it is almost worn out. I felt, therefore, that it was up to me to stay on top of the situation and make sure that we got a spacecraft at all, and then try to reassure the engineers that if it satisfied the pilot who had to fly it, it ought to satisfy them. In order to convince them of my case, of course, I had to know the spacecraft myself, inside out."[11]

The assigned Redstone rocket was slated for delivery to the Cape on 22 June, but as that date neared it became obvious that this particular booster would not be ready in time, so another was substituted. This arrived on a cargo aircraft from Huntsville on

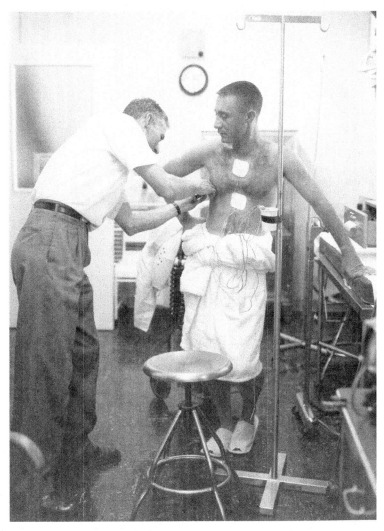

Suit technician Joe Schmitt prepares Grissom for an SEDR test. (Photo: NASA)

the scheduled date. Grissom, in his usual meticulous way, wanted to be part of this. When he got to Patrick AFB at Cape Canaveral the crew had already unloaded the Redstone and placed it on a trailer for the slow ride to the launch pad 15 miles away. "I joined the caravan, and when we reached the pad I got out and walked alongside the Redstone as it pulled in. I guess I looked a little eager, for Paul Donnelly, the capsule test conductor, spoke up.

"'Don't worry, Gus,' he said, 'they're not going to shoot it without you.'"[12]

The MR-4 Redstone rocket arrives at the Cape. (Photo: NASA)

FINAL PREPARATIONS

In time Grissom basically became a living, integral part of Spacecraft No. 11, as the component that no mechanism could replace. His confidence had grown to the point that he knew his vehicle as well as, or better than, any high-performance aircraft he had flown in the past.

"Actually, during the final weeks and days before the launching of MR-4 … I felt really good," Grissom recalled in *We Seven*. "We kept spotting problems, as we knew we would. But there were very few of them, considering the state of the art, and the simulations we went through for practice went very well. If anything was building up inside me, it was that I was anxious. I kept wanting to go tomorrow, and I guess I got slightly impatient whenever some technician came up with a new modification in the system that might have caused a long delay if we had accepted it. The only thing I was afraid of was that something might happen to prevent *me* from making the flight."[13]

On Friday, 23 June, Redstone launch vehicle MRLV-8 was installed on Launch Pad 5 for a mission expected in mid-July. Technicians began conducting extensive check-outs of the 69-foot rocket before mating it with the one-ton Mercury capsule, checking and rechecking to ensure the booster was ready for the upcoming flight. Once the launch pad crews had completed their final inspections and systems tests, *Liberty Bell 7* would be moved from Hangar S to the launch pad and mated with the Redstone. Three Redstone rockets, including the one on the launch pad, remained for the planned Mercury suborbital flights. If the NASA schedule held, the last two launchings would take place in August and September.

Grissom conducts systems checks inside *Liberty Bell 7*. (Photo: NASA)

The MR-4 Redstone being raised to vertical at Launch Complex 5. (Photo: NASA)

Preparing to mate *Liberty Bell 7* with the Redstone booster. (Photo: NASA)

As Grissom recalled, it was a happy day for him when the booster and spacecraft were finally being spliced together. Incredibly, however, he was almost barred from being on the pad to observe the process. "I had locked my hard hat in the office and forgotten the key, and no one is allowed near an active gantry without a special hard hat to protect his head. Someone finally loaned me one, and I made it just in time."[14]

Over the next three days, further compatibility tests involving thousands of parts would take place in order to ensure that all systems involving the spacecraft and booster worked together. George Baldwin served as a manufacturing foreman with McDonnell, overseeing the launch pad crew. He spent his days at the Cape in preparing for the Mercury launches. He still recalls that time with great fondness, as he said in 2011. "My experience with it was wonderful because of the camaraderie and willingness of the workers and engineers [and] because everyone had one goal in mind. It was a time when we had Sputnik going over top of our heads, and [President Kennedy] setting the goal of going to the Moon within the decade. It was an absolutely exciting time."[15]

As final preflight operations proceeded on schedule at the Cape, NASA personnel began manning stations on Bermuda and Grand Bahama Island in readiness to track

Mating the Mercury spacecraft with the Redstone booster. (Photo courtesy Kansas Cosmosphere and Space Center)

the MR-4 flight. At sea, ships and aircraft of the Mercury recovery force were either on station or moving into position, ready to pluck Grissom and his capsule from the sea.

GRISSOM NAMED TO FLIGHT

On Thursday, 29 June, NASA said there were tentative plans to launch the MR-4 mission on a downrange suborbital flight from Cape Canaveral during the week of 16 July. They said the name of the chosen astronaut would be revealed on Monday, but everyone expected the pilot to be John Glenn, with Grissom on the third flight. NASA also revealed that it had originally planned to launch the second Redstone rocket and capsule just six weeks after Shepard's flight. However it was forced to delay these plans because of changes made in the capsule in response to Shepard's recommendations. The bulletin stated that it was expected that the astronaut on this flight would have fewer tasks to perform, and have more time available for Earth observation.

On 10 July Grissom was an interested spectator at a demonstration of the explosive hatch, held at the Cape Canaveral Missile Test Annex. It was identical to the one installed in *Liberty Bell 7*. Harry Lutz from McDonnell was the pyrotechnic engineer in charge of the test. Once everything was ready, the spectators moved off a safe distance and the explosive material was detonated by means of pulling on a long lanyard attached to the T-shaped exterior initiator. The hatch blew as planned and the test was deemed a complete success. Grissom went away satisfied this system would work well on the actual flight.

Technicians prepare to test the explosive hatch system. (Photo courtesy of Kansas Cosmosphere and Space Center)

The hatch is ready to be tested. (Photo courtesy of Kansas Cosmosphere and Space Center)

McDonnell's Ralph Gendielle (at right) examines a hatch bolt after the test. Partly obscured behind him is John Yardley. (Photo courtesy Kansas Cosmosphere and Space Center)

The following day NASA reported that the flight would likely be set for 18 July, but the space agency declined any further comment on the identity of the astronaut. They said they would reveal the name of the chosen astronaut on Monday, 24 hours ahead of the launch. Then, on 15 July, and to the surprise of many, Robert Gilruth finally confirmed to reporters that Grissom would be the prime pilot for the MR-4 mission, with Glenn acting as his backup. It was also announced that Grissom had chosen the call-sign *Liberty Bell 7* for his spacecraft, and an engineer had stenciled the name in white paint onto the side of the capsule.

"One of the less vital problems I had was figuring out a name and an insignia for the capsule," Grissom later revealed. "As the pilot, I had the prerogative of thinking up a name. I decided on Liberty Bell, because the capsule does resemble a bell. John Glenn felt that the symbolic number 'seven' should appear on all our capsules – in honor of the team – so this was added. Then one of the engineers got the bright idea

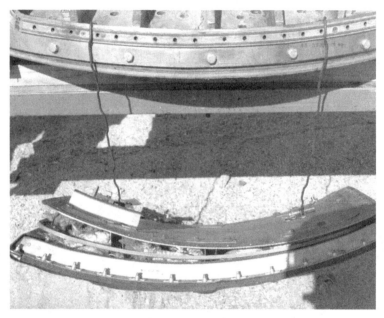

The hatch, still attached as planned to the door sill, rests on the ground. (Photo courtesy Kansas Cosmosphere and Space Center)

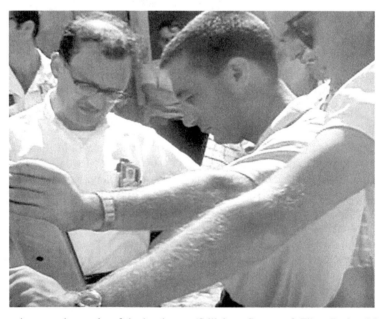

Grissom inspects the results of the hatch test. (Still from Spacecraft Films Project Mercury film set)

that we ought to dress Liberty Bell up by painting a crack on it just like the crack on the real one. No one seemed quite sure what the crack looked like, so we copied it from the 'tails' side of a fifty-cent piece."[16]

Interestingly, Jerry Ter Horst of the North American Newspaper Alliance later claimed that the choice of Grissom as prime pilot for the MR-4 mission over John Glenn was the result of what he called "the Air Force factor." As Ter Horst stated, this hint of inter-service rivalry for space flight honors came to him "unofficially and privately" from NASA sources.

"According to the sources, Capt. Grissom was destined to be chosen over Marine Lt. Col. Glenn of Arlington, Virginia, because of the Air Force's primary interest in space and because the Navy already has an astronaut in Cdr. Alan B. Shepard, Jr., who made the May 5 flight.

"'Gus' Grissom, 35, is one of three astronauts from the Air Force. Col. Glenn, 40, is the only Marine astronaut. The Marine Corps, however, is actually a part of the Navy. Three of the astronauts, including Cdr. Shepard, are Navy pilots. None of the seven is an Army man.

"'All things being equal in their personal readiness and training for this shot, the choice of Grissom over Glenn was indicated because of the Air Force factor,' one source said privately. 'Imagine how the Air Force would feel if it missed out on the first two flights,' said another.

"There has been no hint of inter-service rivalry among the astronauts, who have lived and worked together as an inseparable team for two years. Officially, NASA has played down their basic membership in the armed forces because of the civilian nature of this country's space program. Col. Glenn had been presumed in line for the first trip, but the nod went to Cdr. Shepard. Col. Glenn became his 'backup' man. Col. Glenn then became the seeded choice for the second manned flight into space. Capt. Grissom, from Mitchell, Indiana, apparently has been the secret NASA choice for several weeks – with full knowledge of the other astronauts.

"Sources also said that astronaut No. 3 will not necessarily be Col. Glenn. The 'top three' selection was only for the first two flights, it was said, and all seven astronauts will be in the running for subsequent flights – including Cdr. Shepard and Capt. Grissom."[17]

On the Friday evening before flight week, Grissom attended a meeting on the preparedness of the Redstone and discussed some minor problems in loading liquid oxygen into the rocket. At the end of the meeting he admonished everyone to leave well enough alone by saying, "Don't anybody fiddle with it over the weekend." He then flew home to spend a few final hours with his family before returning to the Cape on Sunday.

A GIRDLE AROUND THE WORLD

Human space flight was in its infancy in mid-1961, and whilst many things had been taken into consideration for the comfort and safety of the astronauts, many unusual and unexpected issues would tend to crop up that needed a little additional thought and initiative. One such problem occurred during the flight of Alan Shepard, and it fell to the astronauts' nurse Dee O'Hara to make a secret shopping trip prior to the flight of Gus Grissom.

When O'Hara first met the Mercury astronauts at the Cape after taking up her duties in Hangar S as the astronauts' nurse, she understandably felt quite intimidated by the seven pilots and the aura surrounding them, even before the first space shot. But as she got to know the men she not only grew comfortable with them, and vice-versa, she also formed a lasting bond with them and their families. It was a bond based on friendship and mutual trust. She got on well with all of them, but admitted it took quite some time to connect with one of the seven, Gus Grissom.

"The only one I didn't get to know right away or feel really close to was Gus, for some reason. But Gus was very quiet. I mean, it took time with Gus. The others were … it was almost, I don't know what to say … just that it didn't take very long till we reached the stage where we were comfortable. They knew me and, you know, it just evolved. But Gus, you had to kind of work at that one."

O'Hara was asked if this was a trust thing on the part of Grissom, as she was in the medical profession, which all pilots shunned as much as possible.

"I don't know, it very well may have been," she ventured, "because back then there were no women in the [space] business. There was no one in that hangar, except there was an occasional … there were one or two secretaries, and I was not wanted at all by the management of NASA. It was [as if] they didn't want me out there. It was a total, total male world. You know it was all engineers, and they flat out did not want any nurse up there – let alone a female. I didn't know a lot of this had gone on; I didn't know I was not wanted at that point – I had no idea … so I was pretty ignorant of the facts.

"But with Gus, he was just comfortable with other men and other pilots, and maybe it was the medical thing. I have no idea. I guess I never figured that out … but it took a long time, and so many months for him to look me in the eye or ask me for something, whereas the others it was, well it just came very natural. But not with Gus, and I don't know why."

Shortly before Grissom's flight, O'Hara was asked to help resolve a very delicate matter, based on a pre-launch problem involving Alan Shepard. In retrospect, no one seems to have considered the fact that a lengthy delay might have an adverse effect on an astronaut's bladder, and, as a result, Shepard finally had to urinate in his space suit.

"Well, there were so many delays," O'Hara recalled of that day. "It hadn't been a problem until there was a launch delay after delay after delay. Finally poor Alan had been out there for what, four, six hours, and in the end they just said, 'Hey, go ahead and urinate in your suit.' And Alan always laughed and said he was the first wetback in space. But then … he had no other choice, and so then they tried to come up with a solution for Gus. That's when I got sent on a mission for a girdle!"

It was not exactly a top-secret task, but Dee O'Hara was asked if she would make her way into Cocoa Beach and quietly locate an item of women's apparel that would soon make its way into space.

"At that time they had these god-awful latex girdles, panty girdles, and I had to go in [a shop] and find one that would fit Gus. And I did." Once there, an unknowing store assistant asked if she could help, and O'Hara said, 'Well, I need a girdle for a friend.'

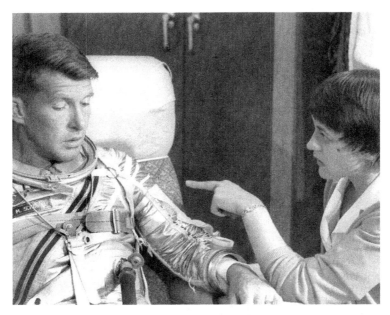

Dee O'Hara with Mercury astronaut Wally Schirra. (Photo: NASA)

"The store assistant asked 'Well, what size is she?' And I said, 'Well gee, I don't know.' She said, 'Well they come in all sizes – you have to have some idea some idea what size she wears.' So I picked out something that I thought might – Boy, if they only knew. So I picked out something I thought that might fit, and fortunately I think it did. Anyway, they used that, fitted out with a condom in order to … in case he needed it."[18]

READY TO GO

Three days before the planned MR-4 launch, Grissom participated in a mission dry run on Launch Pad 5. By this time he had moved out of the Holiday Inn in Cocoa Beach and was staying full-time in crew quarters at Hangar S. The simulated flight involved securing the side hatch, purging the crew cabin with oxygen, and rolling the launch gantry away.

"I showed up early at Pad 5 for the simulated flight which would be the final practice mission in the capsule before the launch," he recalled. "It went fairly well, but I was kept so busy handling communications checks that I fell slightly behind in the count. All of the sequences in the countdown took place in the right order, but some of them came off a little late. Then the second of the three retro-rockets, which are programmed to fire at 5-second intervals, went off two and a half seconds early. So we had to check into that."[19]

Joe Schmitt suits up Grissom for the simulated flight on 16 July. (Photo: NASA)

Apart from these problems, the test was successfully completed and everything seemed to be in order for the actual launch.

On Sunday, 16 July – a day described by Grissom as "a fairly lazy day" – he and Glenn entered the final phase of preparation for Tuesday morning's expected space shot. For the three days before the planned launch date they both lived in the crew quarters of Hangar S, where they had a comfortable bed, a television set and radio, reading material and complete privacy. It also provided them with isolation from any possible carriers of infectious disease organisms.

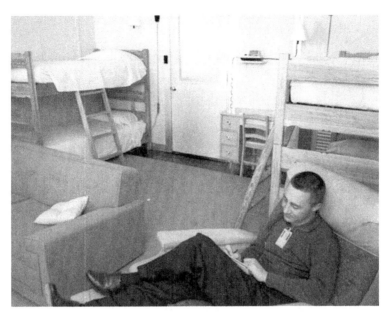

Grissom relaxes in the astronauts' crew quarters in Hangar S. (Photo: NASA)

The previous day, the two astronauts had begun a low-residue, high-energy diet to reduce the possibility of excretion and provide quick-burning reserve strength during the flight. They ate in a special ready room at the Cape, their meals being made by a personal chef whose sole duty during that time was to prepare their food. According to the astronauts' flight surgeon Dr. William Douglas, the menu was prepared "by Miss Beatrice Finklestein of the Aeromedical Medical Laboratory, Aeronautical Systems Division, U.S. Air Force Systems Command." As Dr. Douglas recorded, the chef prepared identical meals at each feeding; two went to the pilot and backup pilot, while several were given to other people "so that an epidemiological study can be facilitated if necessary." An extra serving was kept in a refrigerator for 24 hours "for study in the event that the pilot develops a gastrointestinal illness during this period or subsequently." Furthermore, no coffee was permitted during the 24-hour period preceding either suborbital flight because of its tendency to inhibit sleep, and none was permitted for breakfast on launch day because of its diuretic properties.[20]

For Grissom and Glenn, breakfast consisted of four ounces of strained orange juice, half a cup of cooked Cream of Wheat hot cereal, two or three slices of crisp Canadian bacon, two scrambled or boiled eggs, white toast, butter and strawberry jelly. For lunch they had broiled chicken, baby-food type peas, bread without crust, cottage cheese salad, ice tea, and sugar cookies. Dinner was broiled potato without skin, baby food vegetable and sherbet.

On Sunday afternoon the two astronauts traveled to nearby Patrick AFB to begin their final preflight physical examinations, following which Grissom relaxed with a little surf casting on the beach of the Cape missile center.

The Pad 5 blockhouse at Cape Canaveral. (Photo: NASA)

A rocket stands ready to fly. (Photo: NASA)

On Monday, 17 July, NASA announced that everything – including the weather – appeared to be set for the space shot. Lt. Col. John A. ('Shorty') Powers of NASA's Public Affairs Office informed newsmen, "As of this time all elements are 'A-OK' for this mission." The first use of the term 'A-OK' was once mistakenly accredited to Alan Shepard but it was actually Powers who introduced the phrase to the public, using it during his live broadcast of Shepard's flight. Powers was a decorated former transport pilot who flew in World War II and Korea before serving as an information officer for the Air Research Development Command. He became the public face of NASA and the Mercury program and quickly earned for himself the sobriquet "the voice of the astronauts." He knew that NASA engineers used the 'A-OK' term in radio transmissions tests because the sharper sound of the letter A cut through static better than the letter O. Liking the military snap of the phrase, Powers borrowed it for his mission broadcasts.

Powers also confirmed before an audience of 200 newspaper, radio and television reporters that Grissom remained the prime pilot for the MR-4 flight, with Glenn as his backup. He added that Shepard would be CapCom for the flight, backed up by Scott Carpenter. Slayton would be at the communications post in the blockhouse, maintaining radio contact with Grissom in the capsule before the launch. Schirra would be Slayton's backup for this role. Gordon Cooper would observe the launch while flying high over the Cape in an F-106 jet.

Grissom and Glenn were asked to go to bed early on the evening before launch, but only as a suggestion, not a strict recommendation. As well, and unlike Shepard, Grissom would optimize his launch morning routine by shaving and bathing before retiring. By 5:00 p.m. on Monday afternoon he was fast asleep in crew quarters. At 10:30 he was woken by Bill Douglas and informed that the launch had been called off owing to low-hanging clouds spawned by a Caribbean weather front. Grissom, whose ability to relax under pressure helped him win MR-4, accepted this news with a yawn and went back to sleep.

Fortunately the Redstone rocket had not been loaded with liquid oxygen. This meant officials could reset the shot for the same time Wednesday without having to purge out the rocket. According to a NASA spokesman, it also meant that the first segment of the 12 hour-long countdown, which was completed Monday, would not have to be repeated. "It's locked up," the spokesman said, adding that the remaining hours of the countdown procedure would be picked up starting about 11:30 that evening.

SECOND ATTEMPT

The weather at Cape Canaveral on Tuesday morning was overcast but the engineers and scientists once again busied themselves preparing the Redstone. By this time the beaches and nearby roads were lined with hundreds of tourists anxiously waiting for the launch spectacle. Grissom and Glenn took advantage of the postponement to get in a little running and to study a map of geographical features that Grissom might be able to see while looking down on Earth from an altitude of 115 miles. Grissom then grabbed a fishing pole, strode to the beach and caught a four-pound bass – which he threw back. Afterwards he returned to the business of getting ready for his flight the next day.

Confident that the weather would be fine by launch time, NASA decided Tuesday night to go ahead with plans to send Grissom into space the next morning. Informed sources said rain squalls which had drenched the rocket firing center late in the day would move to the north by early morning, clearing the skies.

Grissom said he watched an episode of *The Life and Legend of Wyatt Earp* on television that evening before going to bed at nine o'clock.

"I slept like a brick for four hours and woke up wondering what time it was and what the weather was like. Just then Bill Douglas came in and sat down on the bed. He just sat there for a few moments, but when he saw me looking at him he said simply, 'Well, get up.'"[21]

After being told the one remaining weather hazard was an area of rain moving across the Gulf of Mexico, which the weather people expected would break up on the west coast of Florida, Grissom was told that the count had been pushed ahead by an hour to try and beat the weather. The launch was now provisionally set for 7:00 a.m.

"The schedule allotted me 30 minutes for breakfast, another 30 minutes for a short physical [examination], 10 minutes to fasten on the electronic sensors which would report my pulse, temperature and breathing rate back to the Control Center, 30 minutes to put on my pressure suit, then 30 minutes to get into the van and ride out to the pad."[22]

Apparently someone forgot to pass on word about the earlier launch time to the cook, as breakfast was not ready at 1:45 a.m. as intended under the revised schedule. A decision was made to conduct the physical exam first, and then tackle breakfast. Grissom donned his bathrobe and entered the medical room where Bill Douglas was waiting. As he subsequently recalled, there was nothing unusual about the preflight examination except that Douglas was concerned about the astronaut's blood pressure count.

"It can't be this low," he said over and over again. Grissom, calm and composed, suggested with a laugh that he could boost it up a bit for the records if he wanted, but Douglas remained amazed at the unexpectedly low reading.

"I think you ought to be just a *little* bit excited," he said with a smile and a shake of his head. Next, Grissom had a short session with NASA's consultant psychiatrist George Ruff, who wanted to explore the astronaut's thoughts.

"He made me recite my feelings, and then we played some little games with words and numbers – to make sure I was completely sane, I guess."[23]

Grissom asked that his wife, Betty, and other members of his family be advised by telephone of the advance in the launch schedule. Breakfast was then shared with Glenn and Scott Carpenter, along with NASA Operations Director Walt Williams, following which Grissom had biomedical sensors attached to his body. To ensure that these were correctly positioned, Grissom (like the other astronauts) had a two-millimeter-diameter tattooed dot at each of the four electrode sites. Then he suited up with the assistance of suit technician Joe Schmitt, and was ready for a pressure check of the suit at 3:10 a.m. Once this had been completed, Grissom boarded the big white transfer van outside Hangar S at 4:15 for the ride over to Launch Pad 5.

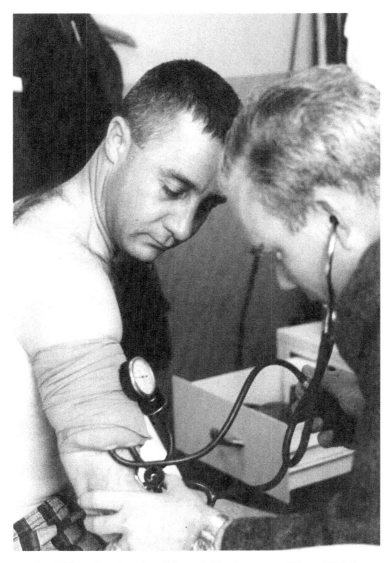

Dr. William Douglas takes Grissom's blood pressure. (Photo: NASA)

At the pad, everyone remained inside the air-conditioned vehicle as Grissom waited for word to step out and make his way to the gantry elevator. At one stage Deke Slayton entered the van to give Grissom a final weather briefing, which was not all that positive because cirrus clouds were moving in and thickening as they approached the Cape. However the local weather reports suggested the skies might be clear enough at launch time to proceed. At 5:00 a.m., word arrived that Grissom could make his way over to the gantry elevator, for the ride up to the spacecraft on the third gantry level, some 65 feet above the ground. The skies were clear over the launch area.

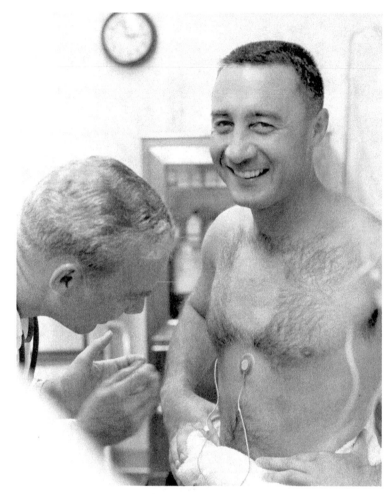

Totally relaxed, Grissom laughs as the biomedical sensors are glued to his chest by Dr. Douglas. (Photo: NASA)

"I stepped out of the van, took a quick look at the tall, white Redstone, and headed for the elevator," Grissom later said. "Just then all the men working around the pad started to applaud. I must admit this choked me up a little. It was a darn fine feeling, as I looked down and saw them staring up at me, that I had all these people pulling for me."[24]

On arrival at the third level Grissom walked across the gantry platform and at 5:38 a.m., after some preliminary procedures, he began to squeeze his 155-pound frame into the capsule with the assistance of John Glenn. Once inside, Joe Schmitt hooked him up to the capsule's air and communications systems and strapped him tightly into the contour couch. After shaking hands with some of the gantry crew, Grissom started to thank Glenn, who unexpectedly handed him a note which read "Have a smooth apogee,

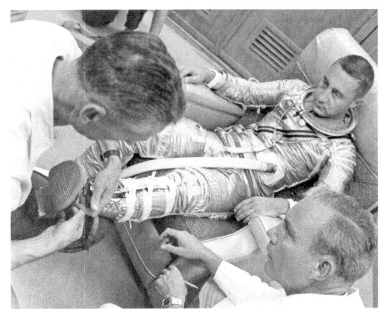

Bill Douglas looks on as suit technician Joe Schmitt laces Grissom's boots. (Photo: NASA)

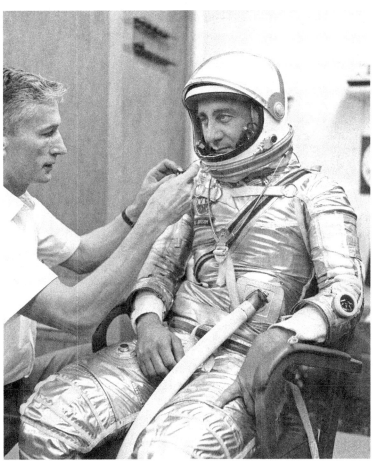

Joe Schmitt prepares Grissom's helmet. (Photo: NASA)

Ready to board the transfer van, Grissom shakes hands with Operations Director Walt Williams as Scott Carpenter (left) looks on. (Photo: NASA)

The NASA transfer van departs Hangar S for Launch Pad 5. (Photo: NASA)

Gus, and *do good work*. See you at GBI [Grand Bahama Island]." Grissom laughed at the reference to his own brief motto delivered at the Convair plant, and after he had shaken Glenn's hand the hatch was bolted in place. He was now alone inside *Liberty*

As Grissom waits at the pad in the transfer van, Deke Slayton gives him the latest weather forecast. (Photo: NASA)

Bell 7, with very little do as he waited for the firing order. Noticing that his window was a little smeared with fingerprints from well-wishers who had pressed up against the glass, he reported this to Guenter Wendt and was assured that it would be wiped clean. One technician who happened to overhear this exchange joked with Grissom they would be sure to install windscreen wipers on the capsule before its next launch.

Eventually the gantry was withdrawn and the Redstone stood alone, poised for launch and pointed at the sky.

The countdown clicked along to within 10 minutes, 30 seconds of firing. Then NASA officials called a halt to study the weather situation. The 'hold' dragged on and the count was recycled to 30 minutes. Finally at 9:00 a.m., two hours after the scheduled time, the reluctant but necessary decision was made to postpone the shot because the heavy, high-level cloud cover was too dense to permit camera coverage of the first critical moments of rocket flight and photographic tracking of the rocket. The launch was rescheduled for 6:00 a.m., Friday. A bad guess on the weather had led to the scrub. A two-day postponement was required because the Redstone had to be purged of its fuels, dried out, cleaned, and checked for contamination.

Grissom had sweated out 3 hours, 57 minutes in the cramped capsule, but as he climbed out was still able to muster a weak smile. "I was disappointed, however, after spending four hours in the couch. And I did not look forward to spending another 48 hours on the Cape.

"But I felt sure we would get it off the next time around. And we did."[25]

Grissom heads for the gantry elevator. (Photo: NASA)

Grissom clambers into a waiting automobile following the launch scrub for his return to Hangar S. (Photo courtesy Kansas Cosmosphere and Space Center)

References

1. The *Owosso Argus-Press* newspaper, Michigan, article "Their parents have Mixed Feelings," issue 22 February 1961, Pg 1
2. Grissom, Betty and Henry Still, *Starfall*, Thomas Y. Crowell Company, New York, NY, 1974
3. Carpenter, Malcolm S., Cooper, L. Gordon, Jr., Glenn, John H. Jr., Grissom, Virgil I., Schirra, Walter M., Jr., Shepard, Alan B. Jr., and Slayton, Donald K., *We Seven*, Simon and Schuster Inc., New York, NY, 1962, p. 269
4. *Aviation Week & Space Technology* magazine, article by Marvin Miles, "Marine Stands Out as Astronauts Choice," issue 12 March 1961
5. Grissom, Virgil, *Gemini: A Personal Account of Man's Venture into Space*, The Macmillan Company/World Book Encyclopedia Science Service, New York, NY, 1968, pp. 35/36
6. Glenn, John with Nick Taylor, *John Glenn: A Memoir*, Bantam Books, New York, NY, 1999, p. 178
7. *Results of the Second U.S. Manned Suborbital Space Flight, July 21, 1961.* Excerpt from "Pilot's Flight Report (Virgil I. Grissom)," p. 50, NASA Manned Spacecraft Center, Houston, TX, 1961
8. Carpenter, Malcolm S., Cooper, L. Gordon, Jr., Glenn, John H. Jr., Grissom, Virgil I., Schirra, Walter M., Jr., Shepard, Alan B. Jr., and Slayton, Donald K., *We Seven*, Simon and Schuster Inc., New York, NY, 1962, p. 268
9. Grissom, Betty and Henry Still, *Starfall*, Thomas Y. Crowell Company, New York, NY, 1974
10. *Postlaunch Memorandum Report for Mercury-Redstone No. 4 (MR-4)*, Section 7.0 (Pilot Activities). NASA Space Task Group, Florida, 6 August 1961.
11. Carpenter, Malcolm S., Cooper, L. Gordon, Jr., Glenn, John H. Jr., Grissom, Virgil I., Schirra, Walter M., Jr., Shepard, Alan B. Jr., and Slayton, Donald K., *We Seven*, Simon and Schuster Inc., New York, NY, 1962, p. 270
12. *Stet*
13. *Stet*
14. *Stet*
15. George Baldwin interview with Krystal Shetler, *Times-Mail News*, Indiana, July 2011
16. Carpenter, Malcolm S., Cooper, L. Gordon, Jr., Glenn, John H. Jr., Grissom, Virgil I., Schirra, Walter M., Jr., Shepard, Alan B. Jr., and Slayton, Donald K., *We Seven*, Simon and Schuster Inc., New York, NY, 1962, p. 280
17. Ter Horst, J.F., article, "Grissom's Choice Laid To 'Air Force Factor,'" from *Washington Star* newspaper, 20 July 1961
18. O'Hara, Dee, interview with Colin Burgess and Francis French, San Diego, CA, 18 January 2003
19. Carpenter, Malcolm S., Cooper, L. Gordon, Jr., Glenn, John H. Jr., Grissom, Virgil I., Schirra, Walter M., Jr., Shepard, Alan B. Jr., and Slayton, Donald K., *We Seven*, Simon and Schuster Inc., New York, NY, 1962, p. 281
20. Madrigal, Alexis C., article for The Atlantic Monthly Group, "13 Little Things NASA Did to Get alan Shepard Ready for Space," 2013. Online at *http://www.theatlantic. com/technology/archive/2013/08/13-little-things-nasa-did-to-get-alan-shepard-ready-for-space/278500*

21. Carpenter, Malcolm S., Cooper, L. Gordon, Jr., Glenn, John H. Jr., Grissom, Virgil I., Schirra, Walter M., Jr., Shepard, Alan B. Jr., and Slayton, Donald K., *We Seven*, Simon and Schuster Inc., New York, NY, 1962, p. 283
22. *Stet*
23. *Stet*
24. *Stet*
25. *Stet*

4

The flight of *Liberty Bell 7*

It was Friday, 21 July 1961, a morning that was almost a carbon copy of Wednesday prior to scrubbing the launch due to poor weather conditions. For the second time in less than 48 hours, Gus Grissom removed his protective overshoes and was carefully assisted into *Liberty Bell 7* at 3:58 a.m. (EST), hoping to keep his delayed date with history.

A MORNING FILLED WITH OPTIMISM

The U.S. Weather Bureau meteorologists had once again been keeping a close eye on the weather, and especially the areas of low pressure. They kept NASA's flight director informed about conditions not only around the Cape, but also in the landing area. A weather briefing had been held at 2:30 a.m. that morning, and the forecast was positive. Space agency officials noted patches of cloud high above the Cape, but were optimistic the weather would stay good enough to permit the launch.

Everyone who had spoken with Grissom that morning said he, too, was optimistic about his chances and was in excellent spirits. This time, however, there seemed to be a little added urgency attached to the pre-flight launch preparations; it was almost as if the scientists and technicians were unsure if their luck would hold along with the weather.

Grissom was awakened by Bill Douglas at 1:05 a.m. after a little less than four hours of sleep and given the good news that the weather looked good for a launch. Twenty minutes later he sat down with Douglas and Scott Carpenter for breakfast. There was no delay with the meal this time, which was a virtual duplicate of the one he had on Wednesday prior to being driven to the launch pad for the flight that never was.

This time, as Grissom later noted, the buildup to launch time was proceeding well. At 1:55 Bill Douglas began a last full-scale physical to ensure the astronaut had not contracted any last-minute difficulties making him unfit for what would be an often grueling experience. There was another brief session with psychiatrist George

C. Burgess, *Liberty Bell 7: The Suborbital Mercury Flight of Virgil I. Grissom*, Springer Praxis Books,
DOI 10.1007/978-3-319-04391-3_4, © Springer International Publishing Switzerland 2014

Ruff, who found there was no alarming level of anxiety in the astronaut. In fact, for Grissom this was almost part of yet another routine procedure that he had endured many times throughout his training and flight preparation. Ruff easily passed Grissom as mentally fit and ready to go. The biomedical sensors which would monitor the astronaut's heartbeat, respiration and body temperature during the brief flight were attached at 2:25.

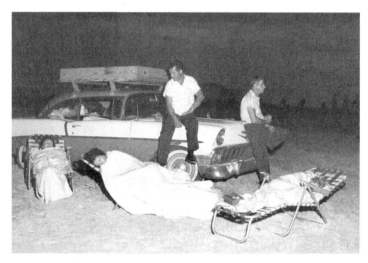

Everywhere along the beaches, people were camped out hoping to witness the historic launch. (Photo: NASA)

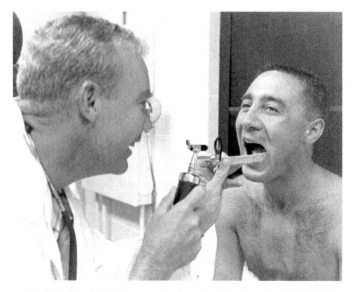

Dr. Douglas gives Grissom a final medical check. (Photo: NASA)

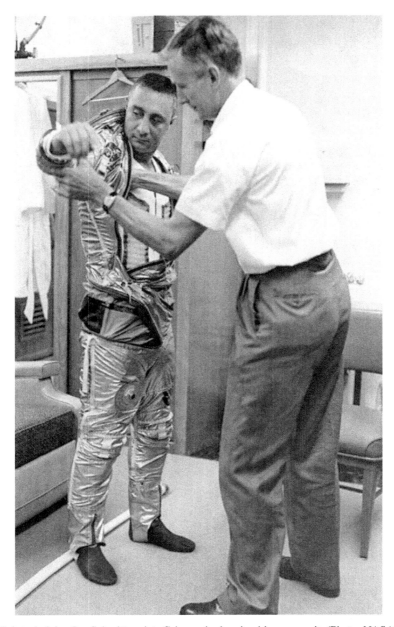

Suit technician Joe Schmitt assists Grissom in donning his space suit. (Photo: NASA)

By 2:55, with the able assistance of Joe Schmitt, Grissom had worked his way into the close-fitting, rubberized and silver-coated space suit. Within ten minutes the suit had been inflated and checked for any possible leaks. No problems were encountered.

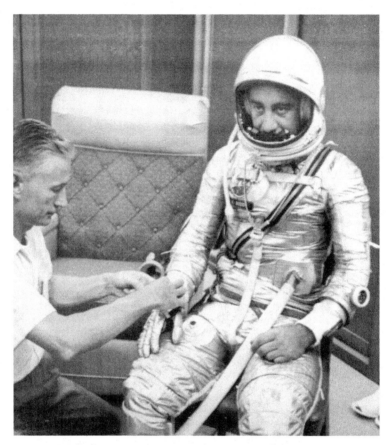

Grissom sits quietly as Schmitt adjusts a glove for him. (Photo: NASA)

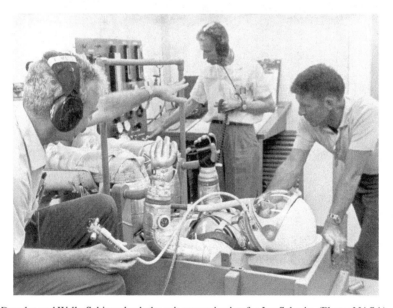

Bill Douglas and Wally Schirra check the suit pressurization for Joe Schmitt. (Photo: NASA)

This time, Deke Slayton visited Grissom at his Hangar S quarters to brief him on the weather, the state of the rocket, and the preparedness of the capsule. Normally this briefing would have been carried out *en route* to the launch pad, but everyone was upbeat about the launch going on time. Grissom then hefted his portable air conditioner, which was known as the "Black Box" and, followed by Bill Douglas, made his way down the stairs to the exit door on the ground floor. Although his mouth was covered by the lower part of his helmet, he could be seen smiling through the open face plate at an assembly of about 60 space agency and air force personnel, photographers, and other spectators. He twice waved his left hand to acknowledge greetings. One NASA official called out, "Good luck," and Grissom responded with an airy "Hi."

Gus Grissom exits Hangar S accompanied by Dr. Douglas. (Photo: NASA)

Riding in the transfer van at a sedate 15 miles an hour, Grissom reached the pad, three miles away, at 3:51 a.m. Two days earlier he had spent nearly an hour inside the van undergoing final briefings, but this time the door swung open in only a few minutes and Grissom, still clutching his air conditioner, cautiously descended the four steps and made his way to the gantry elevator several strides away.

Apart from a quick glance up at the towering Redstone rocket, Grissom looked straight ahead as he walked to the elevator in the peculiar bow-legged gait caused by the tight fit of his suit. This time there was no patter of applause from the helmeted

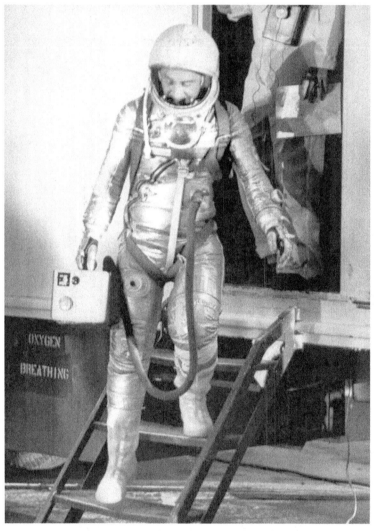

Holding his portable air-conditioning unit, Grissom makes a cautious exit from the transfer van. (Photo: NASA)

While observing activity around the Redstone rocket, Grissom makes his way to the gantry elevator that will carry him to the spacecraft level. (Photo: NASA)

workers on the pad. It was almost as if everyone was holding back their enthusiasm this time until the rocket actually lifted off. Before entering the elevator, however, Grissom exchanged a few light-hearted remarks with some of the men clustered near the elevator cage. Then, with everyone aboard, the elevator rose swiftly up the side of the red steel launch tower. It was a tight squeeze, with the space-suited Grissom, Douglas and other observers making the trip up to the capsule level 65 feet above the ground. The sky above them was dark, but a few stars were evident. Below, the pad area was bathed in dazzling white light from three banks of searchlights, all of them firmly aimed at the rocket.

LAUNCH PREPARATIONS CONTINUE

When Grissom reached the capsule level, he peered briefly down from a window in the green Plexiglas curtain surrounding the capsule, then disappeared from sight of those on the ground. Surrounding the capsule, as usual, were specialists wearing pure white smocks and white skull caps. McDonnell's pad leader, Guenter Wendt, was there as he had been for Alan Shepard two months earlier. "I felt very handsome in my clean white jacket, white baseball cap with the word 'McDonnell' across the front, my headset, and white shirt with a bow-tie," Wendt would later recall.[1]

As Bill Douglas later said of the process of inserting the astronaut into the capsule, "After the pilot climbs into the spacecraft and positions himself in the couch, the pressure-suit technician [Joe Schmitt] attaches the ventilation hoses, the communication line, the biosensor leads and the helmet visor seal hose, and finally he attaches the restraint harness in position but only fastens it loosely. At this point the suit and environmental control system is purged with 100-percent pure oxygen until such time as analysis of the gas in the system shows the oxygen concentration exceeds 95 percent. When the purge of the suit system is completed, the pressure-suit technician tightens the restraint harness; the flight surgeon [Douglas] makes a final inspection of the interior of the spacecraft and of the pilot, and the hatch installation commences. During the insertion procedures, it is the flight surgeon's duty to monitor the suit purge procedure and to stand by to assist the pressure-suit technician or the pilot in any way he can. The final inspection of the pilot by the flight surgeon gives some indication of the pilot's emotional state at the last possible opportunity."[2]

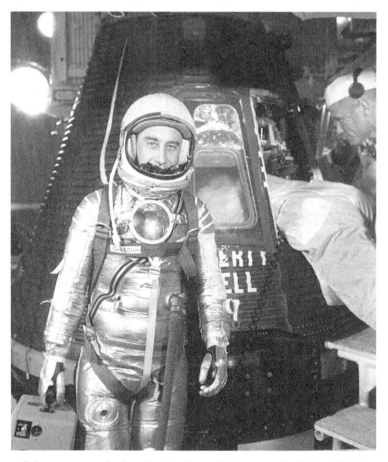

Grissom prepares for insertion into the waiting spacecraft. (Photo: NASA)

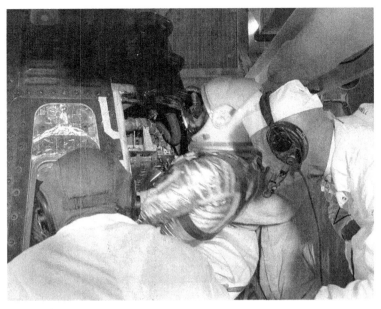

John Glenn assists his fellow astronaut into the tight confines of his spacecraft. (Photo: NASA)

Once Grissom was settled into his form-fitting couch inside *Liberty Bell 7* he was given a reassuring pat on the back by John Glenn, who had once again performed the final checks on the capsule prior to Grissom's arrival. Then Joe Schmitt completed his tasks, strapping the astronaut in and linking him to the onboard communications and ventilation systems before wishing him good luck and withdrawing. Guenter Wendt and his team then took over.

"After final hookups and adjustments, I shook Gus' hand and requested the 'go' to close the hatch," Wendt remarked. "In minutes, my technicians were busy torquing down the 70 hatch bolts. One of the bolts got cross-threaded and we called a halt in the count so that engineering management could assess the situation. It was quickly decided that the one bad bolt would not jeopardize proper function of the hatch and the count resumed."[3] After the flight, Grissom was given that bolt as a souvenir.

Despite the lack of space, Grissom felt quite at home within the tight confines of *Liberty Bell 7*. Countless hours of training – particularly inside the spacecraft – had ensured he was familiar with every switch, system, sound, and countless other facets of the tiny vehicle. "It is good to get into the flight capsule a number of times," he related in his post-flight briefing. "Then, on launch day, you have no feeling of sitting on top of a booster ready for launch. You feel as if you were back in the checkout hangar – this is home, the surroundings are familiar, you are at ease. You cannot achieve this feeling of familiarity in the procedures trainer because there are inevitably many small differences between the simulator and the capsule."[4]

Inside the Mercury Control Center, mission CapCom Alan Shepard prepares for the launch of MR-4. (Photo: NASA)

Meanwhile the tension was slowly building for everyone connected with the flight and the many thousands of people on hand to witness history being made. The Cape and Cocoa Beach areas were slowly coming to life as people woke early, had breakfast, and began gathering on the beaches and every vantage point, nervous but in a state of excited expectation. The news media people started reporting to their assigned pool units, ready to be escorted to the Cape.

COUNTDOWN HOLDS

At 5:45 a.m. the expected liftoff was just 45 minutes away. As the sun rose over the flat, scrub-covered Cape it revealed a pale blue sky with some high, thin, scattered clouds. The weather, it seemed, was also 'go' at this time.

At 5:50 the enclosing launch tower slowly rumbled away from the slender, 83-foot Redstone and everyone's excitement levels began to rise. At the same time the yellow steel "cherry picker" crane moved close to the capsule high atop the rocket. This ungainly-looking device had a cube-like cab at the end of an extended arm to provide a possible emergency evacuation route for Grissom if a serious situation developed before the rocket left the pad.

As the countdown moved beyond 6:00 a.m. – the planned launch time – a brief hold was called at T-30 minutes to allow technicians to turn off the pad searchlights. As it was now daylight and the lights could possibly cause interference with launch-vehicle telemetry, they were no longer needed. By this point, Grissom had been in the capsule for a few minutes beyond two hours.

The "cherry-picker" emergency evacuation device stands ready after the gantry has rolled back. (Photo: NASA)

At 6:25 a.m. there was another hold in the countdown. The reason this time was to allow some clouds to drift out of the way of the tracking cameras. As this might take some time, Scott Carpenter thoughtfully patched a call through to Betty so that she and the boys could enjoy several quiet minutes of conversation with Gus as he waited for the hold to end. Inside their Virginia home, and just like two days earlier, Betty had the astronaut wife's support team of Jo Schirra, Marge Slayton and Rene Carpenter to keep her company, and everyone was crowded around their television set.

"Are you feeling all right?" Betty asked Gus, finding it a little surreal that she was talking to him and watching images of his rocket on the pad a thousand miles away.

"Sure, I'm fine," he responded. "In fact, if they'd stop yacking at me over that darned radio I just might take a nap!"[5]

After a brief conversation with her husband Betty handed the phone over to Scott and Mark, and then he had to sign off.

The sun finally broke through and shortly after 7:00 a.m., with the threatening clouds mostly dissipated, the countdown was resumed with the launch now set for 7:20 a.m. This latest hold had lasted 41 minutes. Grissom, who had been tightly strapped inside the capsule for more than three hours, had spent some of the delay time relaxing with deep breathing exercises and tensing his arms and legs to keep them from getting too stiff.

At 7:10 a.m. the Redstone rocket and the capsule switched over to internal power, meaning that it was now self-sustaining. On the nearby beaches and by the side of roads hordes of undaunted, excited spectators who had endured frustrating days of delay were crowded together with their eyes and binoculars facing the launch pad, but casting an occasional worried glance skyward. With the clouds almost gone, the people began to sense that this might finally be the day the white bird ripped into the sky. Hundreds of weary reporters and cameramen had also taken up their positions ready to finally record and report on the spectacular event.

INTO THE WILD BLUE YONDER

Almost before he knew it, Grissom heard Blockhouse 5 capsule communicator Deke Slayton run the clock down to zero and then call "Ignition!" Grissom felt the launch vehicle begin to vibrate and could hear the engines start. Flames burst from the foot of the rocket.

Moments later the elapsed-time clock started and Alan Shepard, the CapCom in the Mercury Control Center, confirmed liftoff. Grissom quickly performed his next duties. "At that time, I punched the Time Zero Override, started the stopwatch function on the spacecraft clock, and reported that the elapsed-time clock had started."[6] Eight seconds into the launch he almost laughed out loud when he heard Shepard cheekily put on his best imitation of comedian Bill Dana – who had a hilarious routine as a reluctant astronaut named José Jiménez – warning Grissom, "Don't cry too much!"

The Redstone left the launch pad gracefully and drove through a clear patch of blue sky before arching over and heading into the Atlantic target area. The powered ascent proceeded well, and was reported to be very smooth. A low-order vibration became noticeable at around T+50 seconds, but did not cause any interference in communications or degrade Grissom's vision. It quickly dissipated and could no longer be detected by T+70 seconds.

The only problem occurred at this time. One of the carbon jet vanes detached from the Redstone – it can be seen streaking away in footage of the ascent. These vanes, which served as the booster's steering rudder, were mounted in the lower portion of the booster and extended into the rocket's engine exhaust. They were used in conjunction with air rudders to control the Redstone's attitude. During the early part of the ascent

Liftoff of the MR-4 mission. (Photo: NASA)

the Redstone was controlled by the jet vanes, but when the rocket had reached a velocity sufficient for it to become aerodynamically stable, the air rudders took over the control function. The loss of the jet vane at this point did not seem to have any noticeable effect on the stability or function of the booster.

"I looked for a little buffeting as I climbed to 36,000 [feet] and moved through Mach 1, the speed of sound," Grissom later reported. "Al [Shepard] had experienced some difficulty here; his vehicle shook quite a lot and his vision was slightly blurred by the vibrations. But we had made some good fixes. We had improved the

In the Mercury Control Center, Alan Shepard signals the successful launch of the Redstone rocket. (Photo: NASA)

aerodynamic fairings between the capsule and the Redstone, and had put some extra padding around my head. I had no trouble at all, and I could see the instruments very clearly."[7]

Whereas Alan Shepard had been restricted to external observations through a periscope device, Grissom had the benefit of the newly installed centerline window and commented that his vision out of the window was good at all times during the launch.

"As viewed from the pad, the sky was its normal light blue; but as the altitude increased, the sky became a darker and darker blue until approximately two minutes after liftoff, which corresponds to an altitude of approximately 100,000 feet, the sky rapidly changed to an absolute black. At this time, I saw what appeared to be one rather faint star in the center of the window. It was about equal in brightness to Polaris. Later, it was determined that this was the planet Venus."[8]

The Redstone soars away from Launch Pad 5. (Photo: NASA)

As the Redstone continued its ascent, Grissom reported that he was receiving a force of 2.5 G. Then, 142 seconds after liftoff, the Redstone's engine suddenly shut down. Although Grissom reported a slight tumbling sensation and several moments of disorientation, he had experienced similar sensations in centrifuge simulations so he knew what it was and it didn't trouble him. The sensation occurred once again just ten seconds later when the escape tower clamp ring fired and the tower blasted free of the spacecraft. He later said that the explosive separation and firing of the escape tower was quite audible and he could see the escape rocket motor and tower throughout its tail-off burning phase and for some time after that, climbing off to his right.

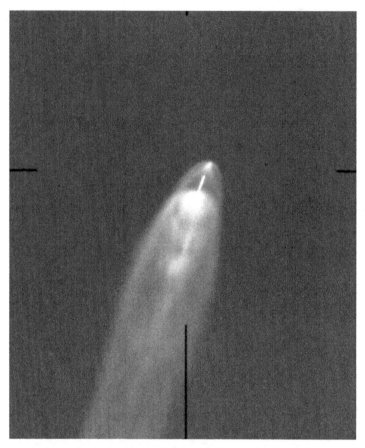

Liberty Bell 7 climbs into the morning sky of the Cape, reaching for space. (Photo: NASA)

Spectators on the beaches below follow the ascent of the Redstone rocket as it streaks ever higher into the sky. (Photo: NASA)

Back in the Mercury Control Center, Wally Schirra lights a celebratory cigar for Alan Shepard. Partly obscured behind and between them is Joe Walker, one of NASA's X-15 pilots. (Photo: NASA)

Grissom was observing the tower when the posigrade rockets of his spacecraft fired on schedule, separating it from the spent Redstone. This was accompanied by "a very audible bang and a definite kick, producing a deceleration of approximately 1 G."[9]

On separation from the booster Grissom pitched forward slightly in reaction, but it was an anticipated sensation. At this point, *Liberty Bell 7* was coasting upward in free flight.

In his later debriefing Grissom said that, like Shepard, he had to make a special effort to notice that he had entered into a weightless condition. His primary cue was a visual one, which became apparent when he noticed stray washers and some trash floating around; there was no other sensation of zero-g.

"Now, I was on my own," he later recorded, as he described the view out of his enlarged window. "Shortly after liftoff I went through a layer of cirrus clouds and broke out into the sun. The sky became blue, and then – quite suddenly and abruptly – it turned black. Al had described it as dark blue. It seemed jet black to me. There was a narrow transition band between the blue and the black – a sort of fuzzy gray area. But it was very thin, and the change from blue to black was extremely vivid. The Earth itself was bright. I had a little trouble identifying land masses because of an extensive layer of clouds that hung over them. Even so, the view back down through the window was fascinating. I could make out brilliant gradations of color – the blue of the water, the white of the beaches and the brown of the land. Later on, when I was weightless and about 100 miles up – almost at the apogee of the flight – I could look down and see Cape Canaveral, sharp and clear. I could even see the buildings. This was the best reference I had for determining my position. I could pick out the Banana River and see the peninsula which runs further south. Then I spotted the south coast of Florida. I saw what must have been West Palm Beach. I never did see Cuba. The high cirrus blotted out everything except the area from about Daytona Beach back inland to Orlando and Lakeland, to Lake Okeechobee and down to the tip of Florida. It was quite a panorama."[10]

RETURN TO EARTH

The flight was going well, and to plan. All too soon the spacecraft's automatic stabilization and control system (ASCS) initiated the turnaround maneuver, placing the blunt end of the capsule forward in preparation for assuming retro-fire attitude. Momentarily, Grissom thought he was tumbling out of control until he realized this was the automatic turnaround procedure, just as he had experienced it on the trainer. As *Liberty Bell 7* swung around, a brilliant shaft of sunlight moved rapidly across his body.

The pitch and yaw axes stabilized with only a moderate amount of overshoot as projected, and was off by approximately 15 degrees when he switched from autopilot to the manual proportional control system. The switchover had occurred ten seconds later than planned in order to allow more time for the ASCS to stabilize the capsule. Having taken over manual control of the spacecraft, Grissom encountered a little difficulty with the attitude controls. They seemed to him to be somewhat sticky and sluggish, and the capsule did not always respond as well as he thought it should.

"I tried to hurry the pitch-up maneuver," was his later observation. "I controlled the roll attitude back within limits, but the view out the window had distracted me, resulting in an overshoot in pitch. This put me behind in my schedule even more. I hit the planned yaw rate but overshot in yaw attitude again. I realized that my time for control maneuvers was up and so I decided at this point to skip the planned roll maneuver, since the roll axis had been exercised during the two previous maneuvers, and [I would] go immediately to the next task."[11]

Grissom wanted to fire the retro-rockets manually whilst simultaneously using the manual controls to maintain the proper attitude. But this was not a critical operation on the suborbital flight, as he was on ballistic trajectory to begin with, and the firing of the retro-rockets was simply an exercise to test their performance. He would later stress that even though he encountered some control problems, and improvements would have to be made, he was confident that he would have been able to handle the situation if he had been in orbit. "That was the main reason I was up there, of course – to find the bugs in the system before we went all the way."[12]

Although slightly behind schedule, Grissom worked hard to get the spacecraft into a good retro-fire attitude. The flight had now reached a much-anticipated point. He had been allocated a full minute for Earth observation. Not long after, the retro-sequence began automatically and Shepard, in the Mercury Control Center, began intoning the countdown to retro-fire. Grissom was still looking out of the window when the count reached zero and, on command, he fired the retro-rockets manually. The thrust buildup was rapid and smooth, but as he continued to gaze out of the window Grissom noticed a definite yaw to the right had begun. He had planned to control the capsule's attitude during retro-fire by using the horizon as a reference, but once the right yaw began he switched his reference to the flight instruments. Once he had scanned these he turned his attention back to the panorama from his window.

"Immediately after retro-fire, Cape Canaveral came into view. It was quite easy to identify. The Banana and Indian Rivers were easy to distinguish and the white beach all along the coast was quite prominent. The colors that were the most prominent were the blue of the ocean, the brownish-green of the interior, and the white in between, which was obviously the beach and surf. I could see the building area on Cape Canaveral. I do not recall being able to distinguish individual buildings, but it was obvious that it was an area where buildings and structures had been erected."[13]

After retro-fire, the retro-jettison switch was placed into the armed position and the maneuvering mode was switched to the rate command control system. Grissom made a rapid check to confirm that the system was working in all axes, and then he switched from the UHF transmitter to the HF transmitter.

"It was a strange sensation when the retros fired," Grissom later recorded for the book, *We Seven*. "Just before they went, I had the distinct feeling that I was moving backwards – which I was. But when they went off, and slowed me down, I definitely felt that I was going the other way. It was an illusion, of course. I had only changed speed, not direction.

"Despite my problems with the controls, I was able to hold the spacecraft steady during the 22 seconds that it took for the three retros to finish their job. Then, right after the retro-pack jettisoned at T plus 6 minutes 7 seconds, and the dead rockets fell away, I looked through the periscope and saw something floating around outside that looked just like a retro-motor. Bits and pieces of the retro-package floated past me a couple of times. It had come loose, just as it was supposed to, and had left the heat shield clean and uncluttered for reentry."[14]

Although Grissom knew the reentry could prove to be tricky, it was nevertheless uneventful. He did report observing what he described as shock waves coming off the capsule. "It looked like smoke or contrail really, but I'm pretty certain it was shocks."[15]

As *Liberty Bell 7* hurtled back through the atmosphere, the g-forces were rapidly building up on Grissom, making his speech and breathing a little more labored. He would report the maximum force he endured was 11.2, but as he had taken as many as 16 G in the centrifuge this was easy to take by comparison. He could also hear a roaring noise, which he presumed was the sound of the capsule's blunt nose forcing its way through the atmosphere.

At T plus 9 minutes 41 seconds the drogue chute came out at the intended altitude of 21,000 feet, right on schedule. Viewing through the centerline window, Grissom saw the canister fall away and the drogue deploy. Next out was the main parachute, 23 seconds later, as planned at 12,300 feet. "I could see the complete chute when it was in the reefed condition," he said in his later report, "and after it opened I could see, out the window, 75 percent of the chute."[16]

One thing he did report at this time was seeing what appeared to be a six-inch triangular-shaped tear in one of the main chute panels. He kept a close watch on this, but it did not seem to grow any larger.

As the spacecraft was slowly swinging and rotating beneath the main parachute, Grissom was able to sense the deployment of the landing bag, intended to soften the spacecraft's impact with the ocean and help stabilize it in the water.

Finally, at T plus 15 minutes 37 seconds, *Liberty Bell 7* splashed into the Atlantic at a rate of 28 feet per second with what Grissom later described as "a good bump." The flight part of the MR-4 mission was at an end.

In the usual competitive spirit of the Mercury astronauts, Gus Grissom would later grin broadly when he discovered his flight had reached an apogee of 102.76 nautical miles, as opposed to 101.24 nautical miles for Shepard's MR-3 mission in May, and it had also lasted 15 seconds longer.

However, unlike the end of Shepard's *Freedom 7* mission, deep and unexpected trouble was about to befall astronaut Gus Grissom and his *Liberty Bell 7* spacecraft.

References

1. Wendt, Guenter and Russell Still, *The Unbroken Chain*, Apogee Books, Ontario, Canada, 2001, p. 39
2. Douglas, William K., M.D., *Flight Surgeon's Report for Mercury-Redstone Missions 3 and 4*, extracted from *Results of the Second Manned U.S. Suborbital Space Flight, July 21 1961*. NASA/Space Task Group, Manned Spacecraft center, Houston, Texas
3. Wendt, Guenter and Russell Still, *The Unbroken Chain*, Apogee Books, Ontario, Canada, 2001, p. 39
4. Voas, Robert B., John J. Van Bockel, Raymond G. Zedekar and Paul W. Back, *Results of Inflight Pilot Performance Studies for the MR-4 Flight*, extracted from *Results of the Second Manned U.S. Suborbital Space Flight, July 21 1961*. NASA/Space Task Group, Manned Spacecraft center, Houston, Texas
5. Grissom, Betty and Henry Still, *Starfall*, Thomas Y. Crowell Company, New York, NY, 1974, p. 96
6. Grissom, Virgil I., *Pilot's Flight Report*, extracted from *Results of the Second Manned U.S. Suborbital Space Flight, July 21 1961*. NASA/Space Task Group, Manned Spacecraft center, Houston, Texas
7. Carpenter, Malcolm S., Cooper, L. Gordon, Jr., Glenn, John H. Jr., Grissom, Virgil I., Schirra, Walter M., Jr., Shepard, Alan B. Jr., and Slayton, Donald K., *We Seven*, Simon and Schuster Inc., New York, NY, 1962, p. 289
8. Grissom, Virgil I., *Pilot's Flight Report*, extracted from *Results of the Second Manned U.S. Suborbital Space Flight, July 21 1961*. NASA/Space Task Group, Manned Spacecraft center, Houston, Texas
9. *Ibid*
10. Carpenter, Malcolm S., Cooper, L. Gordon, Jr., Glenn, John H. Jr., Grissom, Virgil I., Schirra, Walter M., Jr., Shepard, Alan B. Jr., and Slayton, Donald K., *We Seven*, Simon and Schuster Inc., New York, NY, 1962, p. 290

11. Grissom, Virgil I., *Pilot's Flight Report*, extracted from *Results of the Second Manned U.S. Suborbital Space Flight, July 21 1961*. NASA/Space Task Group, Manned Spacecraft center, Houston, Texas

12. Carpenter, Malcolm S., Cooper, L. Gordon, Jr., Glenn, John H. Jr., Grissom, Virgil I., Schirra, Walter M., Jr., Shepard, Alan B. Jr., and Slayton, Donald K., *We Seven*, Simon and Schuster Inc., New York, NY, 1962, p. 292

13. Grissom, Virgil I., *Pilot's Flight Report*, extracted from *Results of the Second Manned U.S. Suborbital Space Flight, July 21 1961*. NASA/Space Task Group, Manned Spacecraft center, Houston, Texas

14. Carpenter, Malcolm S., Cooper, L. Gordon, Jr., Glenn, John H. Jr., Grissom, Virgil I., Schirra, Walter M., Jr., Shepard, Alan B. Jr., and Slayton, Donald K., *We Seven*, Simon and Schuster Inc., New York, NY, 1962, pp. 292-293

15. Voas, Robert B., John J. Van Bockel, Raymond G. Zedekar and Paul W. Back, *Results of Inflight Pilot Performance Studies for the MR-4 Flight*, extracted from *Results of the Second Manned U.S. Suborbital Space Flight, July 21 1961*. NASA/Space Task Group, Manned Spacecraft center, Houston, Texas

16. Grissom, Virgil I., *Pilot's Flight Report*, extracted from *Results of the Second Manned U.S. Suborbital Space Flight, July 21 1961*. NASA/Space Task Group, Manned Spacecraft center, Houston, Texas

5

An astronaut in peril

James D. ('Jim') Lewis, Ph.D., is a former U.S. Marine Corps helicopter pilot, and during his memorable last tour of duty in that service was appointed Mercury Project Officer and became prime recovery pilot for the MR-4 mission.

Lewis continued to serve in the Marine Reserves during his lengthy tenure with NASA, eventually retiring from reserve service in 1983 with the rank of major. His work with NASA and government service would only end in 1999, at which time he was Chief of the Space Human Factors Branch of the space agency.

HELICOPTER RECOVERY PILOT

Jim Lewis was born in Shreveport, Louisiana, on 10 November 1936, which he always proudly states happens to coincide with the birthday of the U.S. Marine Corps. His father had served as a Warrant Officer during World War II, "having enlisted by fudging a bit on his age." After the war the Lewis family relocated to Oklahoma City, where his father had taken on employment with the International Harvester Company. Following several promotions and transfers, the family moved once again to Houston, Texas. Here Jim Lewis attended 6th grade right through to high school, later attending the University of Houston, which meant he could live at home and work various jobs while undertaking his studies. During his junior year he joined the Marine Corps Platoon Leader's program. By attending Platoon Leader's Class (PLC) in the summer between his junior and senior years, he was able to be commissioned a 2nd lieutenant on graduation and receive twelve hours of college credits for the PLC program.

While at Quantico, Virginia, that summer he applied for and was accepted into flight school, which he was scheduled to begin after graduating from college. He admits that he had only chosen the Marine Corps because his father had enlisted in the U.S. Army, and like most young people he wanted to do something different. He served in the Far East after graduation from flight school, spending six months in

C. Burgess, *Liberty Bell 7: The Suborbital Mercury Flight of Virgil I. Grissom*, Springer Praxis Books,   123
DOI 10.1007/978-3-319-04391-3_5, © Springer International Publishing Switzerland 2014

Lewis's Marine buddy Wayne Koons (left) and helicopter co-pilot George Cox flank MR-3 astronaut Cdr. Alan Shepard during a recovery training session. (Photo: Wayne Koons)

Japan and several more on the island of Okinawa, before serving with Marine Light Helicopter Squadron HMR(L) 261 on carriers engaged in supply duties to Vietnam in 1959. However he was not involved in any combat operations.

On his return to the United States he took the advice of Wayne Koons, a friend from flight school, and requested a transfer to the 2nd Marine Air Wing on the east coast at MCAS New River, North Carolina. He opted for this unit because Koons was then involved with the Air Wing in the Mercury capsule recovery program for NASA. Lewis's request was approved, and he later became primary recovery pilot for the MR-4 mission.

"The Marine Corps had been selected [to recover the Mercury spacecraft from the ocean] for several reasons," Lewis told the author. "One was that our helicopters had the payload capacity to lift the capsule. Similar Navy models combined a lot of sonar search equipment that reduced their payloads considerably. In addition, one of the Marine Corps' missions was to deposit heavy external loads in small, tight jungle-type areas surrounded by trees ... a task which required a fairly high degree of precision. While most pilots could accomplish this after training, Marine Corps pilots had been practicing it as part of their normal duties for quite a while."

Another factor in favor of the Marine Corps acting as the recovery force was that they operated a base in Jacksonville, North Carolina, which was reasonably close to Langley AFB, Virginia, where NASA's recently formed Space Task Group (STG) was then located.

Lewis was serving in HMR(L)-262 when he first met Gus Grissom at Langley. The astronaut was visiting the STG for a meeting concerned with the recovery of Mercury spacecraft. Asked how well he got to know Grissom back then, Lewis responded, "I didn't get to know Gus really well ... there was little personal contact at Langley.

The USS *Randolph* (CVS-15) at sea in 1962. (Photo: U.S. Navy)

I didn't get to know Gus really well until I was a Manned Spacecraft Center employee [in Houston, Texas]. I think my impressions were like most. Gus was a serious guy, and the more one had the opportunity to work directly with him, the more one appreciated how good he was. He worked technical problems well, penetrating to the core, and making sure he and all of us took care of any peripheral concerns. In other words, I really appreciated how comprehensive his work ethic was. I imagine that's one of the things that helped him survive his combat missions in Korea."[1]

As primary recovery pilot for the suborbital MR-4 mission, Lewis was assigned to the lead helicopter, a Sikorsky HUS-1 Seahorse of the Marine Medium Transport Squadron, and given the transmission call-sign of *Hunt Club 1*. While in training for the assignment, Lewis and his team practiced for every conceivable scenario, which included the recovery of unmanned capsules from Little Joe booster flights fired out of Wallops Island, Virginia.

As the time grew near for Grissom's suborbital flight, Lewis and his co-pilot John Reinhard from Bloomington, Illinois selected the three best-performing helicopters from their base and flew them to the USS *Randolph* (CVS-15), the prime recovery aircraft carrier. On the morning of the space flight they test flew all three helicopters to ensure they were at peak performance for the recovery effort.

AT THE READY

On the morning of the second launch attempt, as with the postponed liftoff two days earlier, a number of fixed-wing aircraft were flying at high level along the Atlantic missile range in order to assist with the location and recovery of the spacecraft as it broke through the clouds and splashed into the ocean near Grand Bahama Island.

The primary recovery chart for Grissom's mission specifies two P2V Neptune airplanes from the Navy's Patrol Squadron 5 (VP-5) based at NAS Jacksonville, Florida, call-signed that day *Cardfile 5* and *Cardfile 9*. They had SARAH (Search and Rescue and Homing) equipment on board operated by either Navy or Air Force personnel as appropriate, and there was usually a NASA/STG representative. Not shown on the recovery chart was a third P2V call-signed *Cardfile 23*. Piloted by Navy Cdr. Lester Boutte, its assignment was to take up position in the predicted recovery zone, spot the spacecraft as it descended on its parachute and then circle high at high level to observe the recovery efforts. In addition there were two C-54 Douglas Skymasters call-signed *Cardfile 21* and *Cardfile 22*, and a pair of SA-16 Grumman Albatrosses designated *Dumbo 1* and *Dumbo 2*.

The three P2Vs had taken off at staged intervals beginning at 2:00 a.m. In addition to the aircraft flown by Lester Boutte, one was under the command of Lt. Cdr. Edward McCarthy, whose assignment was to fly near the Cape Canaveral launch site, ready to assist in the event of an early booster malfunction over the ocean. A third P2V was operated by Lt. Cdr. Anthony Ruoti, and was stationed downrange from the planned landing site for use in the event of an overshoot.

Cardfile 23 pilot Lester Boutte had been involved in a much-publicized rescue operation some 19 years earlier in November 1942, after a B-17D had been shot down over the Pacific. Boutte, then a radioman aboard a scouting two-man Navy OSTU Kingfisher, had spotted a life raft adrift in the ocean twenty days later, when any hope of finding survivors had all but gone. The survivors, many near death, were rescued and carried to safety – some even strapped to the small aircraft's wings – by the Kingfisher's pilot, Lt. William Eadie, USN, who taxied across the water to a rescue ship. One of those lucky survivors was famed World War I air ace Eddie Rickenbacker.

Also on board *Cardfile 23* as an observer of the MR-4 flight was the STG's Milton Windler. Back then he was a member of the Landing and Recovery Test Section headed by Peter Armitage, one of the Canadian AVRO engineering group that went to work for the newly established space agency NASA. In 1967 he was transferred into Flight Control and served as lead flight director for several Skylab and lunar missions, including Apollo 13 and Apollo 14.

"At the time of MR-4 our Recovery Branch was fairly small; twelve in all, headed by Robert Thompson," Windler reflected. "My job at the time included evaluating, recommending, and testing the Mercury location aids. All of these were activated automatically and required no crew action. This included the SOFAR bombs, HF beacon and the primary aid – the UHF SARAH. This was the same aid as used by the RAF pilots in the Battle of Britain. A very simple, clever scheme. It involved a special receiver and Yagi antennas on the P2V (and other) aircraft. The identical UHF beacon used by the RAF was installed on the Mercury spacecraft.

"We conducted many operational tests and, since I had a lot of experience with these tests, I went out to the primary landing area with the commander of the recovery location aircraft. This was usually (probably always) the senior pilot or aircraft commander for the array. I was there to represent NASA, answer questions and offer advice in the location process, and to provide post mission observations. The aircrews were well

Milton Windler, a later lead flight director with NASA. (Photo: NASA)

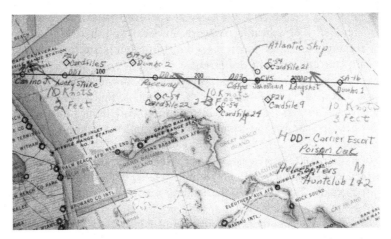

This map, personally annotated by McDonnell engineer Guenter Wendt, shows the position of all the MR-4 recovery force participants. (Photo: Rick Boos)

trained and motivated and really needed little help from me however, except to translate some of the countdown events. NASA had recovery branch personnel with most of the ships as well, especially the [carrier] designated to be the primary recovery ship."[2]

Apart from the USS *Randolph*, the prime recovery carrier, other ships involved in the recovery operation were the destroyers USS *Conway* (DD-507), USS *Cony* (DD-508), USS *Lowry* (DD-770) and USS *Stormes* (DD-780); the oceanic minesweepers USS *Alacrity* (MSO-520) and USS *Exploit* (MSO-440); the tracking ships USNS *Coastal Sentry* (AGM-15) and USNS *Rose Knot* (AGM-14); and the salvage and rescue ship with the appropriately name of USS *Recovery* (ARS-43).

COUNTDOWN TO RECOVERY

When asked how it was communicated to him onboard the *Randolph* that the MR-4 launch had taken place, Jim Lewis vividly recalled the facts.

"My log book shows that we flew two missions that day," he stated, "the first being a checkout flight of just over a half hour. As I recall, our plan was to lift off the carrier at the same time the booster lifted off from Cape Canaveral. We were waiting in the cockpit with engines running and received word that the launch occurred via ship-to-aircraft radio. All that remained was to engage rotors and take off once clearance was granted from flight control.

"Once MR-4 had lifted off, we had about fifteen minutes to get there and begin recovery operations, and I believe the carrier was standing about five miles off the impact area. We flew at about ninety knots, so getting into the primary recovery area quickly was no problem.

"I was initially occupied with observing the sky above, searching for the *Liberty Bell 7* parachute. Beyond that, I was intent on executing the mission procedures and plan. I finally saw the spacecraft on its chutes. I couldn't say what altitude, but it wasn't very high." [3]

With *Liberty Bell 7* now heading towards an ocean splashdown, gently swinging and slowly rotating beneath its main parachute, Grissom heard from the crew of the radio relay airplane call-signed *Card File 23*.

"We are heading directly toward you," the pilot announced, as he observed the bell-shaped spacecraft floating downwards past 3,000 feet. At this time, Jim Lewis aboard rescue helicopter *Hunt Club 1* also established radio communications with Grissom, letting him know that he was positioned about two miles southwest of the projected splashdown site.

As Grissom prepared for splashdown, the protective heat shield at the base of the capsule detached on schedule with an audible "clunk" and dropped about three feet below the spacecraft. This action in the landing sequence also revealed the attached perforated landing bag, which would absorb much of the shock of impact when the spacecraft smacked down on the water. Following splashdown, the bag's next job was to help stabilize the craft by filling up with seawater. It would then act like a sea anchor, to keep the spacecraft upright until it could be hauled out of the water by the recovery helicopter. Salt water would then drain out through air holes in the skirt of the bag.

In the Mercury Control Center, Flight Director Chris Kraft was wary of proclaiming the space flight a success, but as he later wrote he *was* entirely pleased with the way the MR-4 mission had gone. "Grissom was good," he observed. "He handled the maneuvers to perfection, using the three systems of automatic, manual, and rate command, a combination of the two.

"His Earth observations were cogent, and his call-outs during reentry were on time and worry-free. Then he splashed down.

"The radio link between the low-flying aircraft, the helicopters, and mission control was touchy. We only heard part of it. Gus was down and safe … Then next we heard excited voices, too garbled to understand clearly."[4]

PREPARING FOR A PICKUP

At 7:36 a.m., just 15 minutes 37 seconds after lifting off from Cape Canaveral, America's second manned space flight came to an abrupt end as *Liberty Bell 7* plunged into the Atlantic at 28 feet per second. Gus Grissom recalls being a little surprised that the only sensation he felt on splashdown was a mild jolt, which he later equated to impact with the ground after a parachute jump, and "not hard enough to cause discomfort or disorientation."[5]

As expected, the spacecraft heeled over in the choppy seas, and moments later the window was completely underwater. For a few seconds Grissom imagined he was upside-down, but the spacecraft soon began to right itself in the water. He did hear what he later described as a "disconcerting gurgling noise" as *Liberty Bell 7* slowly rolled upright and the recovery section on top of the capsule drew clear of the water. A quick check reassured him that no seawater was entering the spacecraft. His next action was to jettison the reserve parachute by clicking a recovery aids switch. He heard the chute jettison and through his periscope could see the canister drifting away in the choppy water. He reported that he was in good shape whilst switching on a radio beacon and other rescue aids and deploying a sea marker which spread a bright green dye. He then began to complete his final checks, as he later reported in his post-flight debriefing.

"I felt that I was in good condition at this point and started to prepare myself for egress. I had previously opened the faceplate and had disconnected the visor seal while descending on the main parachute. The next moves, in order, were to disconnect the oxygen outlet hose at the helmet, unfasten the helmet from the suit, release the chest strap, release the lap belt and shoulder harness, release the knee straps, disconnect the biomedical sensors, and roll up the neck dam. The neck dam is a rubber diaphragm that is fastened on the exterior of the suit, below the helmet-attaching ring. After the helmet is disconnected, the neck dam is rolled around the ring and up around the neck, similar to a turtleneck sweater. This left me connected to the spacecraft at two points – the oxygen inlet hose which I needed for cooling, and the helmet communications lead."[6]

Next, Grissom radioed the waiting helicopter pilot Jim Lewis. "Okay, *Hunt Club*, give me how much longer it'll be before you get here."

"This is *Hunt Club*," Lewis responded. "We are in orbit now at this time around the capsule."

"Roger," Grissom radioed back. "Give me about another five minutes to [record] these switch positions here, before I give you a call to come in and hook on. Are you ready to come in and hook on any time?"

Aboard *Hunt Club 1* Lewis was prepared for the operation. "Roger. We're ready any time you are."

"Okay; give me about another three or four minutes here to take these switch positions, then I'll be ready for you."[7]

According to plan, Grissom then noted down all the switch positions using a grease pencil. "All switches were left just the way they were at impact, with the exception of the rescue aids, and I recorded these by marking them down on the switch chart ... and then put it back in the map case."[8]

Liberty Bell 7 after splashdown, showing the HF whip antenna fully extended. It had to be cut by the prime helicopter crew prior to attempting to descend and hook onto the spacecraft's recovery loop. (Photo: NASA)

Recording the positions with a grease pen while wearing pressure suit gloves proved difficult, but he completed the task. The suit ventilation was continually causing his neck dam to swell, which he alleviated from time to time by jamming a gloved finger between his neck and the dam to allow a buildup of air to escape.

Once his shut-down duties and checks had been completed, Grissom turned his attention to oft-rehearsed preparations to blow the hatch. "I took the pins off both the top and bottom of the hatch to make sure the wires wouldn't be in the way, and then took the cover off the detonator and put it down toward my feet." He had earlier removed the hand-forged Randall survival knife from its sheath in the door and placed it in his survival pack as he called in the rescue helicopters. He admits that a short while later, as he lay back in his couch waiting for the helicopter to arrive overhead, he contemplated retrieving the knife from his survival pack and keeping it as a souvenir of the flight. Specially made for NASA, only nine were produced: one for each of the astronauts, and two spares. They were one of the strongest knives ever made, fashioned from high-grade Swedish steel, which Gordon Cooper once described as "so sturdy that it can be used like a chisel to cut through steel bolts. You could probably slice your way right through the capsule wall with it if you had to."[9]

Grissom was satisfied that he had completed all of his pre-egress tasks. Once the prime helicopter had hooked up to *Liberty Bell 7* and raised the hatch clear of the water the pilot would confirm that all was in readiness. All Grissom had to do then was punch the detonator plunger with a five-pound blow and the hatch would explode outwards from the spacecraft, allowing him to slide out onto the door sill and wait for the rescue sling to be lowered. It was then just a matter of being winched up into the helicopter. Following this, he and the capsule would be transported together to the waiting carrier, emulating the successful retrieval of Alan Shepard and *Freedom 7* some ten weeks earlier.

Meanwhile, Jim Lewis had responded to Grissom's transmission, advising the astronaut that the rescue sling would be ready for him outside the capsule once the hatch was explosively jettisoned. Each of the Marine rescue helicopters carried a two-man crew; on *Hunt Club 1* Lewis and his co-pilot Lt. John Reinhard were working well together as a team. Despite the noise within his cockpit, Lewis later reported communications with the spacecraft were "normal and excellent, as the system was designed for that acoustic environment, so all was nominal. All our voices were calm. We'd rehearsed these procedures and activities [in Chesapeake Bay] off Langley AFB with the astronauts, and there was no reason not to be calm. That's what good training does. In addition, Gus had a low resonant voice, which was pleasant to hear. It was all very calm and professional. There were Navy aircraft in the area available to provide communication relays in case they were needed and to extend the search area if needed. As I recall, there were Russian trawlers in the general area, but not the actual recovery zone."[10]

ASTRONAUT OVERBOARD!

Recovery procedures now meant that Lewis and his co-pilot Reinhard had the task of cropping off most of the spacecraft's 4.2-meter HF whip antenna so that it would not interfere with the helicopter's main rotor when the craft came down to hook onto the lifting loop in the recovery section of the capsule. The telescopic antenna had been automatically deployed when the spacecraft landed, and was there to provide long-range communications and emit a signal for a contingency recovery in the event of the capsule coming down well away from the planned splashdown zone.

As Grissom had landed in the planned pickup zone the antenna was no longer needed, and the procedure was to cut most of it off before the helicopter moved in. This was accomplished using shears similar a tree pruner, attached to the end of a long pole. The co-pilot's job was to reach down and sever the antenna using this implement. Once he had hooked onto a strong Dacron loop located at the top of *Liberty Bell 7*, the pilot would apply sufficient engine power to hoist the capsule around 18 inches until the base of the hatch was clear of the water. Finally, they would lower the rescue sling to a position just above the spacecraft and relay this information to Grissom. The astronaut would then disconnect his helmet (thereby ending communications), power down the spacecraft, blow the hatch, and wait for the rescue collar to appear. He would then carefully egress through the hatch and insert his arms into the loop at the foot of the sling. Apart from the use of an explosive hatch, this was the same procedure that had worked so well for Shepard.

As he awaited further instructions, Grissom suddenly heard a dull thud. Without warning, the spacecraft hatch was gone – blown off and out into the ocean. "There wasn't any doubt in my mind as to what had happened," he would report in his later debriefing.[11] Grissom looked up in shocked disbelief, not only seeing blue sky, but the unnerving sight of salt water spilling over the bottom of the door sill and into the spacecraft.

Grissom knew from previous experiments that once water reached the lower edge of the hatch opening, a Mercury spacecraft could sink in just ten seconds. "I made just two moves," he later remarked, "both of them instinctive. I tossed off my helmet and then grabbed the right edge of the instrument panel and hoisted myself right out through the hatch."[12] Moments later he was in the water and being thrust away from *Liberty Bell 7* by the fierce rotor wash of the helicopter amid a tumult of noise and a rolling Atlantic swell.

In shock, Grissom tried to swim backwards away from his spacecraft, watching with mounting horror as seawater continued to pour in through the open hatch. Co-pilot John Reinhard was equally surprised as Grissom "swam out of the capsule and swam away."[13]

Grissom then discovered he had become tangled in a dye-marker line that was wrapped around his shoulder. The line was attached to the spacecraft and he knew that if he could not free himself there was the very likely prospect of being dragged down with the spacecraft if it sank. He finally managed to disentangle himself and moved away from the line.

Somewhat comforted by the fact that he was floating well above the water line and treading water in his buoyant space suit, Grissom was more concerned with the pilot's efforts to hook onto *Liberty Bell 7*, which was rapidly settling into the water. "As I got out, I saw the chopper was having trouble hooking onto the capsule. He was frantically fishing for the recovery loop. The recovery compartment was just out of the water at this time and I swam over to help him get his hook through the loop."

By this time the helicopter was directly over the spacecraft, but with all three of its wheels dangerously in the water. "I thought the co-pilot was having difficulty in hooking onto the spacecraft and I swam the four or five feet to give him some help," Grissom later stated. "Actually, he had cut the antenna and hooked the spacecraft in record time." [14]

Those who ever doubted the bravery and tenacity of Gus Grissom have obviously not viewed the film footage of what happened next. Battered by the ferocious force of rotor wash and whipped-up water, his immediate concern was not for himself but for his spacecraft. He was not sure whether the helicopter was fully hooked onto the

As *Hunt Club 1* hovers dangerously low overhead, accompanied by a Navy support helicopter, Grissom checks to see that they have a solid hook onto the spacecraft's recovery loop, which is just above the water. An enhanced view of Grissom from the same photograph shows his courageous efforts to ensure the retrieval of *Liberty Bell 7*. Note his helmet floating in the sea at top left. (Photos: NASA)

Taken from film footage of the recovery efforts, this still frame shows Grissom trying to hold *Liberty Bell 7* upright as *Hunt Club 1*'s wheels dip into the ocean. (Photo from NASA footage)

spacecraft's Dacron recovery loop, which was perilously close to being submerged. He actually reached out and checked this, and then pushed back from the spacecraft, giving John Reinhard a quick double thumbs-up to indicate that they were properly hooked up. At this time, only a few inches of the very top of the spacecraft could be seen above the water. It was an incredible act of sheer bravery and guts which could so easily have cost him his life, and demonstrated that despite the circumstances, Gus Grissom was not lacking in a test pilot's greatest attributes – his coolness and courage under extreme peril.

As he watched, all the time being forced away from the scene by the downdraft, Grissom knew that things were not going well. "The helicopter pulled up and away from me with the spacecraft and I saw the personal sling start down: then the sling was pulled back into the helicopter and it started to move away from me."[15]

Expecting as a result of previous recovery training that the astronaut should stay comfortably afloat in his space suit, and therefore was not in any immediate danger, Jim Lewis concentrated on saving the spacecraft from sinking to the bottom of the sea.

Meanwhile, Grissom suddenly realized that he too was sinking ever deeper in the water which, fortunately, was not all that cold. The neck dam was working well, so that was not the problem. It had been designed and tested by fellow astronaut Wally Schirra specifically to prevent water entering a floating astronaut's space suit, and it probably saved Grissom's life that day. But he realized that in his haste to evacuate from *Liberty Bell 7* he had neglected to lock the midsection oxygen inlet port on his space suit. This was allowing the air in his suit to bleed out and seawater to seep in. With every second that passed he was becoming increasingly heavier and sinking ever lower in the churned-up seawater. He reached down and locked the suit inlet connection to prevent further water penetration.

Hunt Club 1 manages to partially raise the spacecraft as Grissom looks on. In the lower photo the USS *Randolph* can be seen in the distance. (Photos: NASA)

Now struggling hard to stay afloat, Grissom also recalled some souvenir items he had stuffed into the left leg of his suit. These comprised of two rolls of fifty dimes, some miniature models of his spacecraft, and a small wad of dollar bills. "They were added weight I could have done without," he later confessed, even though the weight was actually quite minimal.

A SINKING SPACECRAFT

Unaware that Grissom was struggling to stay afloat in the ocean swells of three to six feet amplitude, Jim Lewis called upon all his flying skills and the performance of his helicopter to attempt to salvage the sinking spacecraft. He and co-pilot Reinhard had seen the hatch blow off prematurely, and as Lewis watched in alarm "it hit the water, skipped once and sank when it hit the second time." He then related what happened next.

"I was not worried about Gus being in the water because we had trained on these procedures at Langley AFB and the Space Task Group and we knew the astronauts floated very well in their suits – they were sealed and had a neck dam at the top to prevent any water ingress. At that point we no longer had communication, so there was no way for any of us to know there was an open port in his suit.

"My last call to Gus before the hatch blew was that I was 'turning base'. That meant I was downwind and had to do a 180-degree turn into the wind and complete the approach over a distance of one hundred feet or more to get there. We saw the hatch blow, which means that we had completed the turn but still hadn't closed the distance."

"My plan at that point was to have my co-pilot cut the HF antenna … and try and snag the capsule before it sank. There was probably a minute or less from the time the hatch blew until the capsule disappeared below the surface.

"I could see Gus in the water, trying to help in the recovery process. He later said he wanted to help my co-pilot make the connection between the aircraft and capsule if he could, so he was close by. It turned out he didn't need to help [and] he did not look like he was in distress during the time I could see him, and he looked intent on doing what had to be done – but never did he look angry. Anger is a wasted emotion at such times, and pilots are trained to be resourceful, efficient, skilled, and to get the job done, whatever it is.

"I had to put the wheels in the water – the aircraft wasn't designed for this – after my co-pilot cut the antenna so he could reach the recovery bale on top of the capsule. By the time he had made the connection between the helicopter recovery line and the capsule recovery loop, the top of *Liberty Bell 7* had actually disappeared below the surface. Once the hookup was made I could no longer see the capsule because it was directly below the aircraft. I began attempting to lift it out of the water at that point, although I knew that the combined weight of the capsule and water was more than the [helicopter's] lifting capacity."[16]

According to rehearsed procedures, Reinhard's next action was to lower a horse-collar hoist for Gus to climb into, so that they could pull him up into the helicopter. Lewis, meanwhile, was hoping that he might be able to raise the capsule sufficiently to allow much of the water to drain out of it and from the landing bag. It might, he reasoned, give him a fighting chance of hauling *Liberty Bell 7* across to the waiting carrier.

"The landing bag was draining fine when we lifted the capsule out of the water," Lewis said. "We even managed to get the capsule out of the water several times." This usually occurred when the capsule was in the trough of a swell, but as the next swell rolled in, which could be up to six feet high, the spacecraft would once again begin to disappear beneath the surface as water gushed into the open hatch, thereby dragging the helicopter back down again. At these times it weighed 1,000 pounds more than the helicopter could normally lift. As Lewis recalls, this was a tense and potentially calamitous situation.

"I was using maximum power at this point, some 2,800 rpm and 56.5 inches of manifold pressure. Shortly after I began this process, I saw the chip detector warning light on the helicopter instrument panel illuminate. This light indicated there were metal filings in the oil system. Our standard operating procedure for this event said that the engine would probably last about five minutes with metal being distributed throughout the engine before it failed. Because of this, I ordered the co-pilot to cease lowering the hoist for Gus and to bring it back up because we had a sick bird, and I didn't want to lose the aircraft with Gus aboard it. Water egress from a helicopter down in the water with rotors turning overhead is neither a risk free nor an easy task.

"I called the backup helicopter, told him I had a chip detector light, and to come in and pick up Gus, and I also said that I'd drag the capsule clear of Gus so he could come in and make the pickup. Dragging it away was not that easy, but we managed to get it clear in a couple of minutes."[17]

The pilot of the backup helicopter was Capt. Phillip Upschulte from Quincy in Illinois. As the engines of Lewis's aircraft strained against the ponderous weight of the submerged capsule while dragging it through the water, Upschulte maneuvered in behind Lewis's craft. Co-pilot Lt. George Cox then lowered a rescue sling for the waving astronaut, who was now some 70 feet away from his sunken spacecraft.

CLOSE TO DEATH

As the drama continued to unfold, it was being captured on film by photographers aboard a second Navy helicopter, whose pilot had been instructed to remain well clear of the recovery area. Their dramatic images caught Grissom wallowing in the ocean swell. Every so often, the powerful down draft from the helicopters would momentarily force him below the surface of the water, and then he would bob up again like a cork.

Donald Harter from Columbus, Ohio recalls that day as one filled with many emotions. He was a DC2 (Damage Control, 2nd Class Petty Officer) temporarily assigned to the rescue/recovery team aboard the USS *Randolph*, and was aboard Navy Sikorsky HSS-1 N No. 52 from Helicopter Anti-Submarine Squadron HS-7 ("The Big Dippers"). Positioned in the helicopter's doorway, he witnessed the drama below from close hand.

"I had trained in astronaut recovery previously," he reflected. "The protocol was to *always* have two helicopters involved in the recovery of the astronaut and the capsule. There was another helicopter in the area filled with photographers, but they had been warned to keep well away from the recovery effort.

John Reinhard begins to lower a horse-collar rescue sling down to Grissom before the crew's attempt to raise him was aborted. (Photo: NASA)

"When Lewis notified that he was returning to the carrier because of possible engine failure, we were dispatched as part of the two-aircraft protocol to assist Upschulte and Cox if necessary. If something unforeseen were to happen, we had recovery gear on board; in fact all the helicopters aboard the carrier had winches on the open side door and safety harnesses."[18]

Taken from a high-level recovery airplane this photo shows Lewis's helicopter trying to raise *Liberty Bell 7* while Upschulte approaches Grissom from behind. They are flanked by a Navy support helicopter and (nearest the camera) the photographic helicopter. (Photo: NASA)

Grissom, now struggling to keep his head above water, noticed with alacrity the photographic helicopter in the distance, and would later recall that "I could see two guys standing in the door with what looked like chest packs strapped around them. A third guy was taking pictures of me through a window." Unbeknownst to them, the photographers might very well have been recording on film the last desperate moments of astronaut Gus Grissom, for as he revealed in *We Seven* he was now in a fierce struggle just to stay afloat. Flailing about in the water, he knew he could not last much longer.

"At this point the waves were leaping over my head," he wrote. "I was floating lower and lower in the water. I had to swim hard just to keep my head up. I thought to myself, 'Well, you've gone right through the whole flight, and now you're going to sink right here in front of all these people.'"[19]

Grissom then saw to his relief that the backup helicopter was moving in towards him, dragging a horse-collar sling across the surface chop of the water. He was still in danger of drowning, however, since the rotor wash of the two rescue helicopters and the weight of his waterlogged space suit meant that he could not reach the sling, now about 20 feet away.

Sensing Grissom's problem, Phil Upschulte flew a little further forward, dragging the horse collar right up to the astronaut, who was also swimming towards the sling. Despite his fatigue, Grissom looked up and saw the familiar face of George Cox leaning out of the doorway. The Marine pilot had earlier been involved in successful ocean recoveries of both Alan Shepard and the chimpanzee Ham. "As soon as I saw Cox, I thought, 'I've got it made,'" Grissom later reflected. Now close to complete physical exhaustion, he was able to grab hold of the lifesaving collar. "I had a hard time getting it on," he reported at his post-flight technical debriefing, "but I finally got into it.

Upschulte and Cox finally retrieve an exhausted Grissom from the ocean. (Photo: NASA)

A few waves were breaking over my head and I was swallowing water." In fact, he was so exhausted that he could do little more than pass his arms through the life-saving collar, not caring that it was on backwards, and hang from it.

By this time Grissom had been swimming or floating in the choppy seas for four or five minutes, "although it seemed like an eternity to me," he reflected later. In a departure from procedure, he was dragged 15 feet through the water prior to being hoisted into the air with water streaming from his space suit until he was able to be assisted into the cabin of the helicopter.[20]

Don Harter, watching from nearby Navy helicopter No. 52 remembers the scene well. "We arrived just before [George] Cox had lowered their harness to Gus. We backed off so as not to prop-wash Gus, and let Cox complete the recovery, but still near enough to give assistance. Gus had to swim for the horse collar, with swells going over his head. The NASA helicopter with the photographers who had been waving at Gus while he was trying to stay alive left and returned to the carrier."[21]

Prior to a waterlogged Grissom being hoisted into Upschulte's helicopter, Jim Lewis had been faced with an incredibly difficult decision.

"I was pointed into the wind at this stage and the backup helicopter was behind me, also pointing into the wind to give added lift, so I could no longer see Gus. But as we had communication between aircraft, the pilot of the backup craft let me know when he had Gus aboard. Gus was safely aboard and on his way to the *Randolph* in less than four minutes after the hatch blew on *Liberty Bell 7*, so you can see that our contingency procedures worked perfectly. This is exactly why we had the backup helicopter close at hand.

"After close to five minutes of pulling maximum power [on my helicopter], the cylinder head temperature began to rise and the engine oil pressure began to drop. I decided to release the capsule, so I could set the helicopter down 'normally' in the water if the engine died. That was accomplished by pulling a trigger in the cockpit that caused the hook to open and release the recovery cable from the helicopter. I couldn't see the line but I could feel the result of a reduced payload. I wondered if we could make it back. I declared an emergency at that point, and proceeded back toward the *Randolph*, and was able to land aboard the carrier."[22]

Once onboard Upschulte's helicopter, Grissom shed the horse-collar, and shook Cox's hand with a heartfelt, "Boy, am I glad to see you." Then he grabbed a nearby Mae West life preserver and slipped it over his head. While this might have been standard over-the-water military procedure, he was simply ensuring he was ready in case anything happened to the craft he was now occupying. Once was enough, he figured: he had no wish to endure any more time splashing around in the Atlantic trying to stay afloat. As he fastened the Mae West he was told that his capsule had been lost. He then spent the duration of his ten-minute flight over to the *Randolph* without speaking, tightly buckled up in the lifejacket.

Meanwhile, on the USS *Randolph*, Capt. Henry ('Harry') E. Cook, Jr. had been made aware of the problem as Lewis's helicopter was straining to lift *Liberty Bell 7* out of the water. "The capsule was filled with sea water and so heavy it just couldn't be lifted," he said, looking back to that time. "I was preparing to launch boats to lash on to it and tow it alongside and then use the ship's boat crane to hoist it aboard."[23]

When word passed around the nearly 1,000 crewmen lining the afterdeck and the superstructure of the *Randolph* that the capsule had sunk, there was a brief chill as those watching the drama out at sea feared Grissom himself might have been lost. But the Navy quickly made it clear that the astronaut was safe. Several hundred sailors, including midshipmen from various universities undergoing reserve officer training, were lined up behind ropes on the flight deck of the carrier. But this time there was less of the joyous whooping and hollering that marked the scene on the USS *Lake Champlain* when Shepard came aboard that carrier. Perhaps this relative quietness was due to the observers silently offering thanks that the astronaut had made it.

Meanwhile, back in the Mercury Control Center, everyone was tense, waiting to find out what was happening. They had heard that Grissom was in the water, but the conversation on Chris Kraft's communication line was becoming confusing.

Anxious crewmembers aboard the USS *Randolph* scan the skies as they wait for news on the fate of the astronaut and his spacecraft. (Photos: NASA)

"We didn't know what had happened to Gus," he later wrote. "After a minute of silence, a second helicopter reported that they were 'attempting to recover the astronaut.'"

"What does that mean?" someone innocently asked Kraft.

At that Kraft snapped, and loudly told everyone to just shut up and listen.

"I remember thinking, *I hope it's not a body they're recovering*," Kraft related. "The next minute dragged on forever. Then we heard it. Gus was aboard the chopper. They were returning to the ship."[24]

SAFELY ABOARD

Jim Lewis nursed his helicopter back to the *Randolph*, later acknowledging he was less concerned about the losing the spacecraft than he was in getting his aircraft and crew back safely to the carrier. After landing, his attention was more focused on the wellbeing of Gus Grissom. Like everyone else, Lewis and Reinhard could only wait anxiously for Upschulte's helicopter to arrive.

As Don Harter told the author, although the Navy helicopter and its crew had not been needed in the rescue of Gus Grissom, the astronaut still had to be transported over to the safety of the waiting aircraft carrier. "After George Cox had raised Gus into their helicopter we escorted them to the carrier, as their cargo was top priority. By this time Lewis had already landed on the carrier."[25]

The second Marine helicopter landed gently on the carrier's deck at 8:01 a.m., just 41 minutes after Grissom's Redstone booster had thundered into the Florida skies. Having removed the lifejacket, but still clutching his gloves, a sodden and bewildered Grissom clambered out of the helicopter as soon as it had shut off its engine. The Associated Press reported his first words as, "Give me something to blow my nose. My head is full of water."

There to greet Grissom and hustle him off to his post-flight debriefing were two military doctors – Navy Cdr. Robert Laning and Army Capt. Jerome Strong. They had both performed the same duty after the recovery of Alan Shepard two months earlier.

Laning and Strong would not only conduct a preliminary medical examination and evaluation of the astronaut, but also guide him in the process of "debriefing," speaking into a tape recorder and giving his immediate recollections of the flight. The debriefing

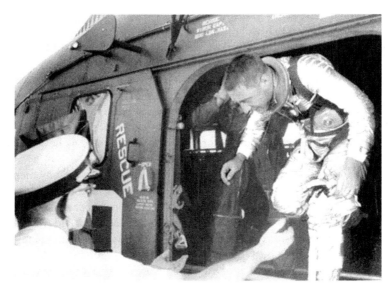

As Grissom is helped from the helicopter by George Cox, Drs. Laning and Strong are ready to assist him onto the *Randolph*'s flight deck. (Photo: NASA)

Still drenched in sea water, Grissom appears to pump his fist at being safely onboard the recovery carrier. (Photo: NASA)

would take place in a quiet, air-conditioned and oak-paneled cabin normally occupied by the *Randolph*'s skipper, Harry Cook. The cabin included a sitting room, a bedroom, a bath, and a small galley.

Before he had a chance to towel off and change out of his water-soaked space suit, Grissom, still dripping water, received a congratulatory telephone call from a relieved President Kennedy.

"Although we were still in the air at the time, we learned the sailors aboard the *Randolph* were deeply sad over the loss of the *Liberty Bell*," Don Harter recalled. "Especially when they saw the helicopters come back without the capsule. However we were proud to be a part of it – a part of history – and the sailors on board cheered as Grissom exited the recovery helicopter, knowing he was safe as he walked across the flight deck while he was being escorted to the debrief."[26]

Despite the passage of more than 50 years, Roger Hiemstra has many memories of the day they welcomed America's second astronaut onboard, although he admits some things have been forgotten with the passage of time. "I remember quite well, though, that very hot and sunny day. We had been primed for the recovery and were kept up to date over the loud speakers. I recall that the flight deck was full of sailors awaiting the recovery. As I knew it could be a long wait and I wanted to get a good location, I had brought a book with me and was sitting on the deck among everyone waiting it out. I do remember a *Look* or *Life* photographer was on board and he came over and took a picture of me reading and asked a couple of questions about what I was doing.

Flanked by Drs. Strong (left) and Laning, Grissom is escorted to the Captain's quarters for a dry change of clothes and a post-flight debriefing. (Photo: NASA)

The scene from the bridge as Grissom is escorted across to the doorway leading to the quarters of the ship's Captain. (Photo courtesy of Otto Preske)

"I had brought along my old 8-mm camera to take some film, but a chief petty officer soon saw me with it and made me put it away. We clearly saw the parachute when it came into view (we were prompted over the loud speakers on the progress and where to look) and watched the splashdown even though it was perhaps a half mile or more from the ship. Even I, as an E3 Yeoman, could soon tell something was wrong. We could see that the helicopter [pilot] had clipped onto the capsule and was struggling to pull it up – at least it certainly looked like a struggle. As we now know water was pouring in. Then there was what appeared to be a cutaway or breakaway and that helicopter struggled to make it back to the carrier. We were told later that the engine had been just about ruined.

"Then there was action around the second helicopter, which was them pulling in Gus. That chopper made it back to the ship just fine and I can clearly see in my mind Gus walking away from the chopper and into the bowels of the ship through a hatch. He passed by within 20-30 feet of where I was sitting. We (the lower enlisted men) never saw him again and I was so saddened to hear later of his death in the [Apollo 1] fire."[27]

Airman Robert ('Bob') Bell was also amongst the crowd of spectators gathered on the flight deck of the *Randolph* for the arrival of astronaut Gus Grissom. "If my memory is correct, I was aft of the number two elevator," he recalled. "So my view wasn't all that great. Everyone was kept back from the landing area for both safety reasons and to shield Grissom. He stepped from the recovery helicopter and nearly ran straight to the hatch leading into the island." Apart from being disappointed at the rapid disappearance of the nation's newest astronaut, Bell believes Grissom was not in a very good mood and still recalls him as being "not at all enlisted-friendly" to the scores of sailors who had gathered on the deck to witness this historic moment. "He may have been upset over the capsule sinking, but he snubbed the sailors on the flight deck and went straight to the officers' area. Nothing like the pickup of John Glenn a few months later," Bell added. "Glenn was a great guy. I still have a slip of paper he signed and dated."[28]

Another excited onlooker was radarman/seaman apprentice Mike Andrew, who was up early in the morning at his station on the *Randolph*'s bridge along with Capt. Cook. After several postponements of the Mercury shot, he was not alone in hoping it would all come together that day.

"At the time, I was a seaman apprentice and the CIC radar representative for Capt. Cook," he explained. "This was my normal station and we were on our fourth mission attempt to retrieve Gus Grissom and his capsule. Lots of volunteers for my station that morning, but *no way* was I going to miss this – I didn't know I had that many friends!

"Spirits remained high and the 'RAND DO – CAN DO' attitude was in high gear throughout the bridge crew, especially since we had the best seats in the house. The clouds seemed less intrusive than the past days and there was a feeling on the ship's island that this was *the day* after three cancelations. For me, at 20 years old, this was a dream of a lifetime, as space flight was the wave of the future and on the carrier's island I had the best seat in the house." There was only one disappointment for Mike Andrew. "I forgot my camera!"

The air of expectancy onboard grew when the sailors learned of the successful launch from the Cape. "Not knowing what was going to happen, all we could do was look towards the heavens and keep an eye on the choppers," Mike Andrew continued. "Each one of us was eager to be the first to sight the capsule. Finally the capsule appears and settles onto a kinda rough Atlantic Ocean. A while later we see a chopper struggling with the capsule. Soon it returns without the capsule which we learned had sunk, but a second chopper had Gus aboard. Our event ended quickly, as once on deck Gus disappeared. I will, however, always remember the smiles and laughter on the bridge that day. High fives hadn't been invented yet. Then, back to work for all. We had a ship to sail."[29]

In watching footage of Grissom's arrival on the *Randolph*, one can understand why he did not think to wave at the surrounding hordes of sailors, or even have time to do so. As he stepped down from the helicopter he was immediately book-ended by Laning and Strong, who quickly ushered the chilled astronaut in his sodden suit across to a nearby hatchway at the foot of the bridge island, asking him questions along the way. Unlike Shepard, who had to traverse a considerable distance of the *Lake Champlain*'s flight deck, and even turn back to retrieve his helmet from inside his spacecraft, a

These two photographs show the recovered parachute canister on the deck of the *Randolph*. (Photo: NASA)

soaked and miserable Grissom was closely escorted just a few feet to the open hatchway and disappeared within seconds of planting his feet on deck. Understandably, despite what he had just accomplished, he was not in the best of moods.

DEAN CONGER RECALLS

Sometime in 1960 the editor of the *National Geographic* magazine offered to lend a photographer to NASA to document the Mercury program. Dean Conger had a long-time interest in aviation and had already documented the X-15 program, so he was selected. In May 1961 he took many of the iconic shots of Alan Shepard that graced the *National Geographic*, newspapers, and other publications. He was also onboard the USS *Randolph* to photograph the recovery of Gus Grissom.

"It was an exciting time," Conger related, looking back. "On the ship I had very little conversation myself with Grissom. I was there to observe and document. Over the course of many months I had met all of the doctors and debriefers who would be in the sick bay; in fact, we became quite friendly and they knew I wouldn't take any compromising pictures."

He did, however, mention that he had two particularly unforgettable memories of photographing Grissom. "One of the first things that happened when he got on the carrier was a phone call from the President … that was the last thing he wanted right then. He still has his space suit on … probably full of sea water. I don't know how it came about, but I also took a picture of him turning his boot upside-down, pouring water out. It was widely published at the time. There were other shots of him after he got cleaned up and the doctors had checked him out. By the time he was ready to fly to Grand Bahama Island he was all smiles."[30]

After the nation's newest hero had been assisted in stripping off his water-filled space suit in the captain's cabin, and had dried off, a Navy officer handed Grissom his helmet, which had been retrieved from the Atlantic. To Grissom's astonishment the officer told him a destroyer crew had plucked it from the sea next to a circling ten-foot shark. He was also presented with a bright orange Navy flight suit to wear. The decorated Air Force test pilot gave a wry smile as he cast his eyes over the suit and could not resist making the wisecrack remark, "I've been trying to get one of these things for years!"

Earlier, when Grissom had stepped down from the rescue helicopter, he had seen a familiar face standing alongside the two doctors. Senior NASA Space Task Group representative Charles Tynan was onboard as the Recovery Team Leader, just as he had been on the USS *Lake Champlain* for the recovery of Alan Shepard two months earlier. And just as on the Shepard recovery, Tynan was wearing his "lucky" yellow patterned shirt.

As Tynan recalled for the author, "I met him at the helicopter with the doctors immediately after the helicopter landed on the carrier, and later approached him in the wardroom. Although we were not supposed to talk with Shepard after his flight, since Grissom's capsule sank I was instructed from MCC [Mercury Control Center] to talk with Gus and see if I could find out why the hatch had blown prematurely. I remember approaching Gus, dressed in a white terry cloth bathrobe, starting to eat breakfast. When I approached, he was spitting out some bad prunes. He cursed and said 'Nothing has gone right today' because the prunes he was starting to eat were no good. Grissom was upset about the capsule sinking but he was adamant about any hint that he blew the hatch, so I didn't learn much about the blown hatch. He didn't want to talk about it. He claimed then, and later, that he did not blow the hatch."[31]

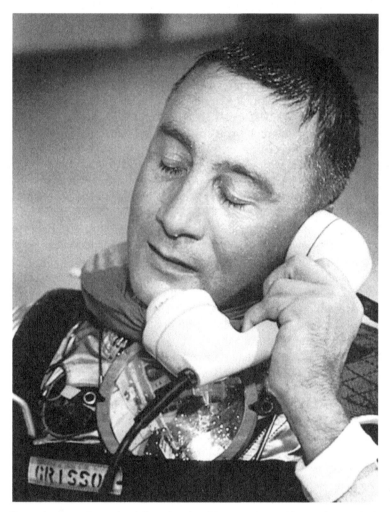

Dean Conger's photo shows the dejected and sodden astronaut taking a phone call from his President. (Photo: Dean Conger/NASA)

During his shipboard debriefing, Grissom paid homage to a life-saving device designed by a fellow astronaut. "Before I end this debriefing," he recorded, "I want to say that I'll ever be grateful to Wally [Schirra] for the work he did on the neck dam. If I hadn't had the neck dam up, I think I would have drowned before anyone could have gotten to me. I just can't get over the fact that the neck dam is what saved me today … So, of course, that's another recommendation: put the neck dam up right away.

"Also, I would recommend that you get the Mylar raft out [of the survival pack] and keep it in your lap before egress even though the chopper is there. I think I was just a little bit over-confident this morning. I saw the choppers were there and so I thought everything was going to be okay. And I almost didn't put the neck dam up … I think

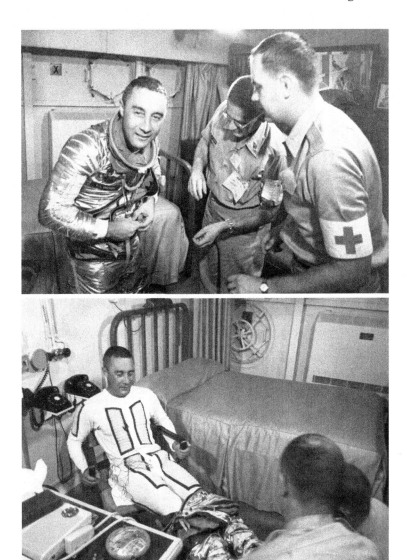

Having spoken to the President, Grissom could finally shed his waterlogged space suit with a little assistance. (Photos: Dean Conger/NASA)

we should plan for a few more emergencies along the recovery line and follow proce-dures exactly as we planned, not get hurried and not get over-confident, either."[32]

There is little doubt that Gus Grissom would have been in dire trouble had he not already been unbuckled with his neck dam rolled up when the hatch blew, giving him barely seconds to make his hurried way out of the sinking spacecraft. The other saving grace was that he was the smallest of the Mercury astronauts. As space historian Rick

NASA's Charles Tynan (colored shirt) was on hand to greet the nation's newest astronaut hero. (Photo: NASA)

Grissom enjoys a hot breakfast despite the presence of some 'suspect' prunes. (Photo: Dean Conger/NASA)

Feeling better after a dry change of clothes and breakfast, Grissom laughs as he pours sea water from one of his space suit boots. (Photo: Dean Conger/NASA)

Boos commented, "When you consider that they were shoehorned in, the angle of the instrument panel relative to the hatch opening, the limited amount of space to be inserted in or to exit from, it was a wonder that he ever was able to exit, especially considering the sink time."[33]

WATERLOGGED, BUT WELL

In a later report for the waiting media, Drs Laning and Strong stated that on arrival, Grissom had immediately received a preliminary physical examination, and blood, urine and other samples had been taken for later analysis. Dr. Strong told reporters that, "Our shipboard examination finds no abnormalities. He is in excellent spirits except that he feels unhappy about the capsule. Both Dr. Robert Laning of the Navy and I are pleased that he is in excellent shape. All tests and observations are within normal limits."[34]

Dr. Laning added that Grissom was in just as good a physical shape as Shepard was after his flight, "except that he was a little more tired after the swim." Laning pointed out that Grissom had swallowed a considerable amount of sea water and was suffering a slight soreness in the throat as a result. He said the astronaut reported he was feeling "a bit shaky" when he first came onboard ship, but soon relaxed and ate a second breakfast that day of orange juice, fried eggs and bacon. The doctor stated that Grissom's pulse rate was 160 when he first came aboard but it shortly dropped to a near normal 100. "He said he had a thrilling ride," Laning told the assembled reporters. "He is feeling well." Laning also said that Grissom asked for a drink of water as soon as he entered the debriefing. "Fresh water, that is," grinned Laning as the reporters jotted down this little gem of information.[35]

Once the initial excitement of Grissom's arrival on the *Randolph* had died down, and while the astronaut was undergoing his preliminary physical examination, Jim Lewis made an official report to the ship's skipper, Harry Cook, on his own part in the recovery effort.

Meanwhile, mechanics inspecting Lewis' helicopter were puzzled, having failed to find anything amiss with the engine. It was later suspected that the chip detector light might have been triggered by a stray metal flake that had somehow worked its way into the engine's oil sump.

Asked what actions he might have taken if the indicator light had not illuminated while he was struggling to raise *Liberty Bell 7* out of the water, Lewis responded, "I would have waited and just hovered with the spacecraft, holding it under water until the recovery carrier arrived. Of course, if that happened, figuring out how to bring the spacecraft aboard the carrier would have been a challenge."[36]

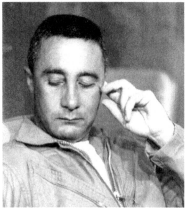

A contemplative Grissom, now wearing his orange flight, shows evident exhaustion as he awaits a flight to Grand Bahama Island. (Photos: Dean Conger/NASA)

Grissom chats to Capt. Harry Cook as he prepares to fly off the USS *Randolph* for his debriefing on Grand Bahama Island. (Photo: Dean Conger/NASA)

Overall, Jim Lewis recalls that everything he experienced onboard the *Randolph* after he landed was "very businesslike and professional because that is how one is trained in the military. Gus understood all that occurred and appreciated the efforts that were made to bring [the capsule] back. While we all would have preferred to have the spacecraft, what resulted, given the circumstances, represented excellent results."[37]

Following Grissom's physical tests and debriefing it was time for the astronaut and the Marine recovery pilots to board a waiting twin-engine S-2F Tracker airplane and proceed to Grand Bahama Island, where the nation's newest hero would undergo a far more extensive medical examination and provide a comprehensive account of his flight, along with his recollections of the post-splashdown dramas. At the island facility the four Marine helicopter pilots would also discuss with Grissom and NASA officials what they had observed from above *Liberty Bell 7*, and attempt to figure out what had led to the loss of the spacecraft.

As he strode over to the flight line with Capt. Cook heading for the S-2F Tracker, Gus was talkative and grinning, and took a moment to wave up to Rear Adm. John E. Clark, who was standing on the bridge. As he reached the airplane, two bells were rung for the attention of all on the carrier and the loud speaker called out, "Captain, United States Air Force departing" – an honor normally only reserved for admirals.

There was another pleasant surprise in store for Grissom courtesy of Capt. Cook as Lewis and Reinhard buckled themselves into two seats in the rear of the airplane. Following a spontaneous request from Cook, and with the agreement of the chief pilot, the Navy co-pilot had happily relinquished his right-hand seat to Grissom.

An S2-F Tracker of VS-26 on the *Randolph*'s elevator. (Photo: VS-26/U.S. Navy)

As Cook shook Grissom's hand he gave him this news. As Cook later observed, while Grissom was clearly pleased to make it safely onto the *Randolph* after his recovery, "he got almost as much thrill … when I invited him to act as co-pilot on the catapult launch of the plane to take him to [Grand Bahama Island]."[38]

The Navy pilot in command on that brief flight was John Barteluce, who, at the age of 90, still hasn't lost his lifetime fascination for aviation. In fact, these days he is the oldest crew member of Coast Guard Auxiliary Flotilla 10-01, whose airplanes fly security patrols each day, extending from the Canadian border to the Manasquan Inlet. "I'm going to go on as long as I can," he told an interviewer in 2012.

Grissom left the carrier by plane for Grand Bahama Island just 77 minutes after stepping onto the deck from the recovery helicopter.

When asked to comment on the events of 21 July 1961, Barteluce told the author, "Needless to say I was happy to be chosen to fly Grissom from our carrier to Grand Bahama Island. Gus was elated that I asked him to be my co-pilot and experience his first and only takeoff from an aircraft carrier.

"It was a fairly short flight and the conversation was mostly about his asking me about my background. I told him that I was a fighter pilot flying F4F Wildcats from the baby carrier USS *Nehenta Bay* [CVE-74] during World War II."

One of Barteluce's most prized possessions is a Dean Conger photograph of him flying alongside Grissom on their way to Grand Bahama Island, which the astronaut later signed for him.[39]

Meanwhile, a press conference was being held at NASA Headquarters at which officials were discussing the loss of the spacecraft. Mercury Operations Director Walt Williams told reporters that the only significant records of the mission lost were two

John Barteluce and his temporary "co-pilot" Gus Grissom. (Photo: Dean Conger/NASA)

color-camera films of Grissom and the instrument panel taken during the flight. They showed the movements of his hands and facial expressions, and might have assisted in improving the positioning of some instruments for later missions.

Robert Gilruth, director of the STG, added that most of the desired information was received and recorded through telemetry. He was quick to point out that despite a dramatic and valiant effort to save the capsule, "the vital part of the cargo – the astronaut – was saved." This sentiment was backed up by another NASA official, who said "We've got only one Gus, but we've got plenty of space capsules."

With questions about the end of the flight still to be answered, Gilruth decided to withhold his planned announcement that the MR-4 flight concluded the suborbital segment of NASA's $400 million man-in-space program, despite there being two suborbital flights listed on the original schedule. He had been expected to say that if everything had worked perfectly, that there was no longer any need to risk the other five astronauts on short but dangerous flights, and the agency would instead start to insert astronauts into orbit using the 300,000 pounds of thrust delivered by the Atlas booster.[40]

Amazingly, the Mercury flight of Gus Grissom had attracted far less attention than that of Alan Shepard just two months earlier. In one serendipitous exercise, the KWK radio station in St. Louis selected telephone numbers at random on the day of the flight and asked thirty people who answered, "Who is Virgil Grissom?" Eleven correctly identified him as the latest astronaut; thirteen said they didn't know; three said he was a disc jockey; one thought he was a radio announcer. Another believed him the former tenant of her apartment, "because I got some of his mail." And one sleepy woman answered, "With this hangover, I couldn't care less!"[41]

References

1. E-mail correspondence James D. Lewis with Colin Burgess, November 2002 to August 2003
2. E-mail correspondence Milton Windler with Colin Burgess, 23 August 2013
3. E-mail correspondence James D. Lewis with Colin Burgess, November 2002 to August 2003
4. Kraft, Chris and James Schefter, *Flight: My Life in Mission Control*, Dutton/Penguin Group, New York, NY, 2001, p. 145
5. Grissom, Virgil I., *Pilot's Flight Report*, from NASA document *Results of the Second U.S. Manned Suborbital Space Flight, July 21, 1961*. NASA, MSC Houston, 1961
6. *Ibid*
7. Grissom, Betty and Russell Still, *Starfall*, Thomas Y. Crowell Company, New York, NY, 1974, p. 100
8. Extract from post-flight debriefing of Virgil Grissom aboard USS *Randolph*, 21 July 1961
9. Carpenter, Malcolm S., Cooper, L. Gordon, Jr., Glenn, John H. Jr., Grissom, Virgil I., Schirra, Walter M., Jr., Shepard, Alan B. Jr., and Slayton, Donald K., *We Seven*, Simon and Schuster Inc., New York, NY, 1962, p. 111
10. E-mail correspondence James D. Lewis with Colin Burgess, November 2002 to August 2003
11. Grissom, Virgil I., *Pilot's Flight Report*, from NASA document *Results of the Second U.S. Manned Suborbital Space Flight, July 21, 1961*. NASA, MSC Houston, 1961
12. Carpenter, Malcolm S., Cooper, L. Gordon, Jr., Glenn, John H. Jr., Grissom, Virgil I., Schirra, Walter M., Jr., Shepard, Alan B. Jr., and Slayton, Donald K., *We Seven*, Simon and Schuster Inc., New York, NY, 1962, p. 295
13. *The Bonham Daily Favorite* newspaper, Texas, "Grissom Barely Got Out as 'Bell' Sank." issue Friday 21 July 1961, p. 1
14. Grissom, Virgil I., *Pilot's Flight Report*, from NASA document *Results of the Second U.S. Manned Suborbital Space Flight, July 21, 1961*. NASA, MSC Houston, 1961
15. *Ibid*
16. E-mail correspondence James D. Lewis with Colin Burgess, November 2002 to August 2003
17. *Ibid*
18. E-mail correspondence Donald Harter with Colin Burgess, 18 June 2013 to 6 July 2013
19. Carpenter, Malcolm S., Cooper, L. Gordon, Jr., Glenn, John H. Jr., Grissom, Virgil I., Schirra, Walter M., Jr., Shepard, Alan B. Jr., and Slayton, Donald K., *We Seven*, Simon and Schuster Inc., New York, NY, 1962, p. 296
20. Extract from post-flight debriefing of Virgil Grissom aboard USS *Randolph*, 21 July 1961
21. E-mail correspondence Donald Harter with Colin Burgess, 18 June 2013 to 6 July 2013

22. E-mail correspondence James D. Lewis with Colin Burgess, November 2002 to August 2003
23. Letter from Capt. Harry E. Cook, Jr. to Mr. Joe Wicks, Gainesville, Georgia, dated 18 April 1977
24. Kraft, Chris and James Schefter, *Flight: My Life in Mission Control*, Dutton/Penguin Group, New York, NY, 2001, p. 146
25. E-mail correspondence Donald Harter with Colin Burgess, 18 June 2013 to 6 July 2013
26. *Ibid*
27. E-mail correspondence Roger Hiemstra with Colin Burgess, 16 May 2013
28. E-mail correspondence Robert Bell with Colin Burgess, 20-21 June 2013
29. E-mail correspondence Mike Andrew with Colin Burgess, 21 June to 7 August 2013
30. E-mail correspondence Dean Conger with Colin Burgess, 23 June 2013
31. E-mail correspondence Charles Tynan with Colin Burgess, 3 January 2013 and 7 July 2013
32. Extract from post-flight debriefing of Virgil Grissom aboard USS *Randolph*, 21 July 1961
33. E-mail correspondence Rick Boos with Colin Burgess, 19 July 2013
34. *Milwaukee Sentinel* newspaper, Wisconsin, issue 22 July 1961, pp. 1-3
35. *Ibid*
36. Swanson, Glen E., "Liberty Bell 7, This is Hunt Club I" article from *Quest: The History of Spaceflight Quarterly*, issue Vol. 7, No. 4, Spring 2000
37. E-mail correspondence James D. Lewis with Colin Burgess, November 2002 to August 2003
38. Letter from Capt. Harry E. Cook, Jr. to Mr. Joe Wicks, Gainesville, Georgia, dated 18 April 1977
39. Bautista, Justo, article "Veteran Who Flew Missions in Three Wars Still Patrolling the Skies," *The Record* newspaper, Hackensack, NJ, 2 August 2012
40. Blakeslee, Alton, article "Space Travelers Chart New Steps," *The Free Lance-Star* newspaper, Fredericksburg, Virginia, 22 July 1961, p. 1
41. *Ibid*

6

One program ends, another begins

Mercury astronaut Gordon Cooper always seemed to live on the edge. Not just the edge of space and adventure, but in testing the patience of his NASA bosses. He loved flying, and his enthusiasm was never more evident than when he finally heard that his pal Gus Grissom had successfully completed his suborbital trip into space.

Cooper was flying an F-106 chase plane over the Cape that day, and he wanted to show the officials below how he felt about his astronaut colleague's safe return from the perils of space. When information was passed to him that the MR-4 mission had been a success he barreled across the Cape, over the heads of newsmen assembled at the press platform, then swung around for a second pass over the area. This time around he performed a slow victory roll, leading NASA's somewhat bemused Public Affairs spokesman Lt. Col. 'Shorty' Powers to announce, "In case there is any doubt in anyone's mind, that was a fellow astronaut who just came by in that F-106, celebrating."[1]

HOMETOWN HERO

Close on 1,000 miles away from the Cape, in Mitchell, Indiana, Gus Grissom's railway signalman father told reporters he had felt mostly fear – "pride ran second" – as his son flew into space and returned to a splashdown in the Atlantic Ocean. Dennis Grissom said that at one point of the flight he became so overwhelmed with a flood of differing emotions he could no longer watch the television coverage and walked into the kitchen away from the grainy images of his son's flight. "It was the longest 15 minutes I ever lived through," he revealed. "You wouldn't realize this unless you had a son up there." However his wife Cecile had endured every tense moment.

C. Burgess, *Liberty Bell 7: The Suborbital Mercury Flight of Virgil I. Grissom*, Springer Praxis Books, DOI 10.1007/978-3-319-04391-3_6, © Springer International Publishing Switzerland 2014

Gordon Cooper flying high above the Cape Canaveral launch site. (Photo: Dean Conger/NASA)

The evening before, the Grissoms had gone on a family picnic with around 40 relatives – "mostly my wife's," Dennis Grissom observed with a smile. But their nerves were still on a brittle edge when they arrived home, partly because their son's flight had already been postponed twice, and then they didn't get to sleep until about 1:00 a.m. Their daughter Wilma Beavers from Baltimore and her children, Rhonda, 12, Joan, 10, and Linda, 9, spent the night with them.

"I just kinda got a feeling they will call it off," Dennis Grissom said of his thoughts before he turned in. Then, at 5:10 a.m. the lights suddenly went on in the little white frame house. Moments before their next-door neighbor, Addie Anderson, fearful the Grissoms might oversleep, had telephoned them. However Dennis said he and Cecile were just about to get up anyway. He then rang a service station across the street and asked the attendant to bring him a pack of cigarettes.

At 5:45 a.m. another son, Norman, a printer on the weekly *Mitchell Tribune* newspaper, arrived with his wife and their daughter Beth, 8. She quickly paired up with Linda Beavers, and the two little girls went out onto the front porch to blow soap bubbles as if it was just another day in their lives. Five minutes before launch time the adults called them back inside. All the children sat on the floor, with the adults nervously occupying the chairs and couches. The volume on the television set was turned right up.

At the moment of liftoff the roar of the Redstone reverberated from the television set. Not a word was spoken; everyone was tense as the gleaming white rocket slowly soared into the Florida skies and Col. Powers began describing the flight. When he mentioned that the escape tower had been jettisoned, Dennis Grissom stood up and walked into the kitchen, where he stayed for several minutes.

Norman Grissom shows his support for brother Gus. (Photo: Associated Press)

Once confirmation came through that their son had been safely recovered, the Grissoms made preparations to move outside and answer questions posed by the assembled news reporters and photographers. They knew by now that the spacecraft had been lost, but all they cared about was that Gus had survived his flight into space, been rescued from the sea, and was said to be uninjured and well. Dennis Grissom put on the coat of his good blue suit, while Cecile smoothed her blue print dress and brushed her hair. Nineteen minutes after the recovery the family stepped onto the porch, Cecile standing with one hand on her husband's shoulder. A nervous but excited Dennis Grissom folded his hands in front of himself and rocked back and forth on the balls of his feet. On his tie, he wore a clasp in the shape of a *Liberty Bell* capsule, just like the one their son had flown into space.

Dennis and Cecile Grissom wave at the crowd gathered around their Mitchell home. (Photo: Associated Press)

The town's mayor, Roy Ira, emerged from the crowd carrying his home movie camera and shook Dennis's hand. Then, like a torrent, came the reporters' questions. Inevitably, the first one centered on how they were feeling

"I feel fine," Grissom said.
"I'm a lot more relieved and I'm glad it's over," Cecile added.
Would they like to see Gus make an orbital flight?
"I think 15 minutes is long enough," Dennis replied.
What about a Moon flight?
"Well, yes. If he can do it safely."
"No. Never," said Cecile.

How did Dennis feel when the space capsule sank beneath the sea after their son had hurriedly evacuated it because the blown hatch permitted water to gush in?

"I was proud he was out of it," was the response. "They can get another capsule …"
A reporter asked whether there were any tears during the flight.
"What do you think?" Cecile replied. "What would you do?"

According to newspaper reports from that day, the Grissoms looked drawn and tired after their son had been plucked from the sea, as if they had been mentally guiding Gus all the way. However they soon joined in the post-flight euphoria and took part in a specially prepared parade through the streets of Mitchell just before midday.

They sat in a convertible and waved to everyone as they trailed behind the high school band and the town's fire truck, and anyone else who wanted to fall in. That day it seemed that this little southern Indiana town's entire population of 3,550 was ready to party, and to celebrate America's second successful space flight by one of their own.[2]

CELEBRATIONS CONTINUE

Also facing a barrage of questions from reporters that day was the youngest sibling of the Grissom family, 27-year-old Lowell from St. Louis, Missouri, who worked as a systems analyst for the McDonnell Aircraft Corporation, the firm which had made the *Liberty Bell 7* spacecraft. He had watched his brother's successful shot from their St. Louis living room with his wife Bobette, and said that they had finally been able to relax for the first time in fifteen days. "We're greatly relieved," he stated. "One more postponement was about all we would have needed."

Lowell disclosed that his brother had told him by phone fifteen days earlier that he would be the pilot for the next mission, well before the public announcement of Gus's selection. "I couldn't tell anyone that Gus would be the pilot," he said. He also revealed that some top McDonnell officials knew his brother had been named, "but they weren't talking about it."

Lowell and Bobette said they only slept "on and off" during the night and were up around 5:00 a.m., "long before the alarm went off." He said that once the Redstone rose from the launch pad safely he was confident everything would go well. His wife said, "I was really shaken up when they said they had lost voice contact for a time. I suppose Lowell was too, but we weren't doing much talking during the shot."

Lowell declined firmly, but politely, to permit newsmen and photographers into their suburban apartment during the space shot, but admitted them once his brother was in the recovery area and ready to be hoisted aboard the helicopter. They were obviously worn down by the two postponements.

"If Gus can stand it, so can we," Bobette said.[3]

In Newport News, Virginia, a proud but relieved Betty Moore Grissom said she was "happy" her husband's flight was a success. "But I'm so sorry the capsule was lost," she remarked.

In her memoir *Starfall*, it was revealed that even though Betty knew Gus's craft had been lost she had no idea how close she had come to losing him. "I didn't have time to worry if he was safe," she explained. "The first thing that went through my head was: I hope he didn't do anything wrong. It was going through my mind, *that* probably was how the news people would write it. I knew if he had made a mistake he would never forgive himself. My second worry was now I had to go out and meet the press."

"I've always known it would be a success," she told a dozen newsmen several minutes later on the lawn of her home, perched on the bank of a small lake, perhaps with more confidence than she felt at that moment. Together with their two sons, Scott, 11, and Mark, 7, and with the wives of her husband's fellow astronauts Deke Slayton, Scott Carpenter and Walter Schirra there to support her, Betty had watched as the dramas unfolded on their television set. She had emerged from her home with Scott

Lowell and Bobette Grissom. Lowell was an engineer at the McDonnell Aircraft Corporation where *Liberty Bell 7* was constructed. (Photo: Associated Press)

and Mark shortly after her husband was safely aboard the aircraft carrier USS *Randolph*. Wearing a light blue dress and a blue and white striped jacket, she was smiling and animated throughout the interview. "We achieved a first today – the boys and I talked by telephone to Gus as he lay flat on his back in the capsule before it was launched. He said if we stopped talking he could go to sleep," she laughed.

How did the boys feel about their father's achievement that day?

"Scott clapped his hands when the rocket went up," Betty said.

"And I whistled too," Scott remarked, then added he would have liked to have been with his father on the flight.

Responding to one question, Betty said that "the last two seconds before liftoff" were the most concerning moments for her. Asked if she prayed during the flight, she said, "certainly." She was also asked if she would like for her husband to be the first astronaut to make an orbital flight. "I think I would, because he would," she dutifully replied.

Betty Grissom at their Newport News home with sons Scott (left) and Mark. (Photo: Associated Press)

When asked about the last time she had seen her husband, Betty replied that she last saw him two weeks before the flight, but had talked to him by telephone daily during that time. "I hope he calls me when he reaches Grand Bahama Island," she added.

Did she wish her husband was in a less strenuous occupation?

"I've always left it up to him to decide what to do," was her considered response. Space flights were important she observed, but said she "will leave them to Gus and the boys."

Did her sons wish to follow in their father's footsteps? She said both boys would probably become pilots.

Finishing up the interview, Betty said, "Now I can rest for a few days and get back to normal." She planned to spend the remainder of that day "watching television and answering the telephone."

"And I'll go swimming," Scott chimed in. Betty would later state that her interview with the newsmen was "much worse than watching the flight on television."[4]

THE PROUD STATE OF INDIANA

In a proclamation signed and dated 21 July 1961, Indiana Governor Matthew E. Welsh declared that throughout Indiana the day would be called "Gus Grissom Day." The proclamation read: "The citizens of Indiana are justly proud of their native son, who showed the exceptional courage and technical skill required to venture into the unknown, and Capt. Grissom's name and daring exploits are now a part of the history of man's pioneering efforts to probe into space. Capt. Grissom has thereby brought honor and renown to his home town of Mitchell, Indiana, and to the state of Indiana."[5]

Only a few hours after the United States had sent its second man into space, President Kennedy signed a bill authorizing vastly expanded space projects, including a start toward sending a man to the Moon. He took note of Grissom's flight as he put his signature to the bill, which authorized the space agency to spend $1,784,300.00 in the year ahead. The amount was every cent Kennedy had asked for.

In a brief statement, the president said it was significant that the bill was signed on the day that America's second astronaut made his flight before the eyes of the watching world and with all the hazard that this entailed.

"It is also significant that once again we have demonstrated the technological excellence of this country," the President said, adding, "As our space program continues … it will continue to be this nation's policy to use space for the advancement of all mankind and to make free release of all scientific and technological results."

The bill had been passed only the day before by the House and Senate.[6]

GRAND BAHAMA ISLAND

On his arrival at Grand Bahama Island, some 4 hours and 18 minutes after the MR-4 liftoff from Cape Canaveral, Gus Grissom jumped from the S-2F, shaking the hand of and receiving a warm welcome from NASA Administrator James Webb, there to meet him.

"Congratulations on a wonderful job," Webb remarked. Noting that Grissom had flown as co-pilot on the S-2F, he added with a smile, "I see you're not tired of flying."

"Not a bit," Grissom responded.

With NASA Administrator James Webb (left) looking on, Grissom is greeted by his fellow astronauts Wally Schirra and John Glenn. (Photo: NASA)

With all the greetings completed, Grissom is escorted to the base commander's car for the drive over to the medical facility. (Photo: NASA)

After thanking Webb for being there, Grissom then shook hands and joked for several moments with fellow astronauts John Glenn and Wally Schirra. "You're looking good," Schirra told Grissom. The astronaut was then ushered into a waiting car belonging to Capt. Hugh May, the Air Force commander of the missile tracking station, and driven about a mile away to a specially built medical facility. The clinic had been erected amid a collection of barracks and buildings that were surrounded in turn by scrub pines and palmettos.

It was originally planned that Grissom would remain on the island for two days before returning to Cape Canaveral for a full-on press conference. While there he would be virtually isolated from newsmen and photographers, who were not permitted direct contact with him because the doctors, engineers and psychologists wanted to question the astronaut without having any ideas planted into his head by outside influences. In addition to further debriefings on the flight, more extensive medical checks would be conducted by Dr. William Douglas and a team of four specialists.

Soon after Grissom's arrival it was time for his first physical checkup, some X-rays, and an hour of freely dictating what he recalled of his flight. He interrupted all this at one stage to call his wife Betty in Newport News to let her know he was safe and well. She was relieved to hear from her husband and jokingly told him that she had been told he "got a little bit wet." On a more serious note she asked if he was okay.

"Sure, I'm fine," he replied. "The water was warm and you know I like to swim." In fact, Grissom was a poor swimmer.

Betty also asked with a little trepidation in her voice how the hatch had blown and if he had anything to do with it. He told her "the hatch just blew." After a little more desultory chat about his flight, he ended the conversation on a lighter note

Escorted by Public Affairs Officer Lt. Col. 'Shorty' Powers (with folder), Grissom arrives at the medical facility with a wave to news reporters. (Photo: NASA)

by asking Betty to bring some extra slacks and shirts for him when she met up with him in Florida.[7]

As the astronauts' aerospace nurse, Dee O'Hara was there to assist Dr. Douglas in evaluating the health of the astronaut.

"Gus arrived at GBI to start his debriefing along with Jim Lewis, the recovery helicopter pilot (and all around great guy) relatively soon after his scary recovery," she reflected. "We didn't have a clue as to what he had just experienced, or at least I didn't. He was quite fatigued, as one would imagine. As soon as he arrived, Bill Douglas gave him his post-flight physical exam. There was a lot of activity going on and Gus was anxious to get on with the debriefing. He did slip out for a moment to call Betty and say he was okay. The plan was for Gus to stay for 48 hours, but Dr. Douglas decided he was recovered sufficiently to return to the Cape. NASA officials were anxious to get started on the briefing and learn exactly what happened to cause *Liberty Bell 7* to sink."[8]

That evening, Grissom watched television reports of his flight. He then spent a quiet, relaxing night, briefly attending a celebration held in his honor by the 263 men – mostly civilians – at the island's auxiliary Air Force base, the first tracking station southeast of Cape Canaveral. He finally retired to his hospital bed around 9:00 p.m.

Meanwhile, news reporters on Grand Bahama Island were eagerly hunting out stories concerning the flight and the capsule mishap to fill their newspaper columns. In order to mollify them a little, a press conference was set up for the next morning with the four Marine helicopter pilots available for questions. Grissom would not be attending.

Once it began, Col. Powers told the gathered crush of reporters that the astronaut looked much less fatigued than the day before. Grissom was in "excellent condition, bright and sharp, and anxious to get home." Asked if the astronaut had ever been in any real danger following his flight, Powers lightly responded, "You're in danger if you're in the middle of the ocean in a pressure suit."

Responding to another question concerning whether Grissom had expressed any dissatisfaction with his own performance, Powers said, "He feels like he has done a good job. From everything I've heard, he indicates he is satisfied with the operation."

"We learned some new lessons from this flight," chipped in James Webb. "The evaluation of them has not yet been made."[9]

As Marine pilot John Reinhard later told space flight historian Rick Boos, the questions from the press that day about the helicopter activities started to get quite pointed, and it all began to get a little out of hand. This is, until a cheesed-off Wally Schirra stepped in.

"Wally Schirra was always my favorite astronaut," Reinhard recalled. "I watched him peel the ass off a bunch of reporters on GBI after the flight. They were asking crap ass questions of us after we lost *Liberty Bell*. You know, we weren't press conscious or anything; we were just a bunch of dumb Marines and [Schirra] stepped forward, and boy, he started in about them giving us a red ass about something that we so valiantly tried to do, and ordered them to stop asking such stupid questions.

"God, I thought Wally was going to kill a couple of those reporters! I could just hear the cameras and microphones being shut off. I guess this is over, I thought."[10]

BACK AT THE CAPE

Meanwhile Grissom had risen at around 6:30 a.m. and eaten a large hot breakfast ahead of a peaceful, relaxing morning out of sight of the press before flying back to Patrick AFB at the Cape aboard a C-54 Skymaster transport plane, together with Deke Slayton and other passengers. Once there, he would address a formal press conference on the results and different aspects of the MR-4 mission.

That afternoon the C-54 landed at the air base where hordes of news reporters and officials were waiting to greet the astronaut. Standing in front of an official NASA limousine were the excited wives of the other six astronauts, holding up a huge banner saying "We're Proud of You All." Sadly for Grissom, however, he also landed in the midst of a gathering pall of public blame and suspected human failure which was about to descend upon the nation's newest spaceman.

Everything seemed to be fine to Gus Grissom as he stepped from the plane into a typi-cally hot Florida summer's day, waving and smiling. At the foot of the stairs he hugged and kissed Betty and their two boys before being ushered off to shake hands with a throng of officials and well-wishers. He was then led into a small annex of a huge tent that had been specially erected for the occasion. Here the family enjoyed a few private moments alone before they were driven to the Starlight Motel in Cocoa Beach where Grissom was to change his clothes and prepare himself for the formal press conference.

Once they had arrived at the motel, Betty and the boys were shown to their seats while Gus – filled with apprehension and unease – stood on a platform facing the reporters and photographers as he prepared to answer their questions. He knew one of the main topics would be the loss of *Liberty Bell 7*, and he simply didn't have any satisfactory answers to offer them at that time. He could only explain what he had done and offer strenuous denials that he was somehow at fault.

Grissom's family was there to greet him on his arrival at Patrick AFB. (Photo: UPI)

As NASA Administrator James Webb looks on, Grissom and his family pose for the hordes of photographs at Patrick AFB prior to his press conference. (Photos: NASA)

To resounding applause, NASA Administrator James Webb introduced Grissom by saying, "I could present Capt. Grissom as an aeronaut, a test pilot, a graduate of that school of experimental flying through which over the 58 years since the Wright Brothers flew, a handful of brave men have taken the personal risks necessary to prove in flight the new aircraft ideas and designs which now benefit so many millions through air transport and add so much to our national security. I could present him as one of the seven astronauts. These seven men have devoted almost three years of their lives to providing the pilot element in the Mercury system, utilizing our most advanced science and technology for the purpose of producing machines capable of aiding man in exploring and extending his knowledge of the universe … for the benefit of all mankind." Webb then presented the press-nervous Grissom with the space agency's Distinguished Service Medal for his "outstanding contribution to space technology."

The Grissom family at the Starlight Motel press conference. Astronauts Wally Schirra and Deke Slayton are seated behind Betty Grissom. (Photo: NASA)

NASA Administrator pins the space agency's Distinguished Service Medal on Grissom's lapel as his family looks on. (Photo: Associated Press)

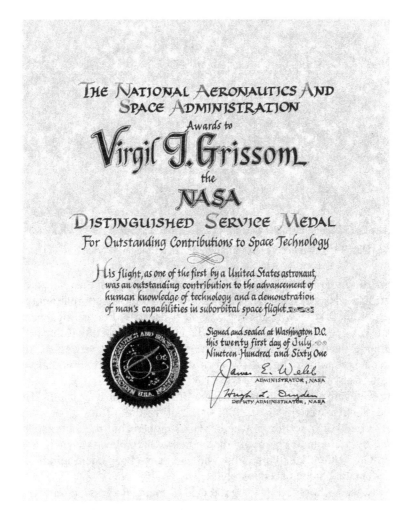

The certificate accompanying Grissom's Distinguished Service Medal. (Photo: NASA)

For his part, Grissom said he was anxious to return to work on the space program and would be back on the job Monday. He repeatedly used the word "fascinating" in describing his first space flight, and said he kept peeking out of his window at the view. He described seeing a band around the Earth that went from light blue to dark blue and then to black, while the horizon appeared to be from 600 to 800 miles from him at the height of his flight.

"Looking out the window – it looked more like a keyhole to me – I could first see blue sky after I went through one little layer of cloud that was floating over the Cape," he said of his ascent into space. "Suddenly the sky went from blue to pitch black. As I looked through my picture window I could see one brilliant star in the middle of the windshield." This "star" would later be determined to be the planet Venus.

Gus Grissom reflects on his flight for the news media. (Photo: NASA)

Grissom said *Liberty Bell 7* experienced much less vibration than Shepard's spacecraft due to some design changes. As the capsule tilted into its "turnaround" position, he said he got his first view of the horizon. "It was really fascinating," he reflected. "The Earth was very bright and very round." He also reported that there was "stuff floating around" inside the capsule during the period of weightlessness, and that "there was in the cockpit some debris – a washer, dirt – the normal debris that you'll find floating around any airplane."[11]

As Betty Grissom and her co-author Henry Still later wrote in *Starfall*: "The reporters skipped quickly over the successful aspects of the flight (they already knew *that*) and probed around the question of whether Grissom had contributed to the loss of the *Liberty Bell* by accidentally bumping the plunger which blew the hatch. Controlling his temper, Gus patiently explained how he had been 'lying there minding my own business' when the hatch unaccountably blew off."

There was, the authors recalled, a good deal of innuendo going around that NASA officials might be somehow trying to cover up a case of pilot error.

"Barroom psychologists whispered the possibility that Gus Grissom was accident-prone. This was typical human perversity, looking for the clay feet of heroes, but it was ironic to those who knew Gus as an outstanding pilot and engineer, a man who devoted endless hours to methodical planning of ways to work around emergency situations."[12]

To a question about whether he ever felt his life was in danger, Grissom answered in his usual honest and considered way, "Well, I was scared a good portion of the time. I guess this is a pretty good indication."

"You were what?" called out one reporter.

"Scared. Okay?" Grissom answered. It was a mistake. His renowned frankness and honesty did not serve him well on this occasion, as the next day's newspapers brought such unjust and sensational headlines such as "Astronaut Admits he was Scared," and "I Was Scared: Grissom." No mention was made of his superhuman struggles against the elements in desperately trying to assist in the retrieval of his sinking spacecraft, an act of incredible bravery which could so easily have cost him his life.

An evidently exhausted Grissom during the press conference, wearing his Project Mercury lapel pin and the Distinguished Service Medal. (Photo: NASA)

James Webb had substituted for the President of the United States in presenting Grissom with the Distinguished Service Medal at the start of the press conference. The reason given was that John Kennedy was dealing with the political aftermath of the Bay of Pigs invasion some three months earlier – although this hadn't precluded a gala White House medal presentation for Alan Shepard the previous month. There would be no White House celebration, no meeting with First Lady Jackie Kennedy for Betty Grissom, no ticker-tape parade through the streets of New York. There was just muted praise and passing recognition, which the astronaut and his wife simply could not comprehend. The guy was a hero; surely they were not blaming him for losing his spacecraft?

Things became even more evident after the press conference when the family was bundled into a car and driven across to Patrick AFB and dropped off at a guesthouse in the base's VIP quarters. They were told that this would ensure a little privacy and security away from the prying eyes of reporters. Meanwhile the other astronauts and their wives were given comfortable accommodation at the Holiday Motel some ten miles north in Cocoa Beach, where they would enjoy post-flight celebration parties. Incredibly, Gus Grissom would have to leave early in the morning; he had been told to report back to work the next day.

"The guesthouse setup didn't suit Betty," Henry Still wrote in *Starfall*. "She did not know how many days they would be there. It was across a busy highway from the Atlantic Ocean beach. Betty's controlled resentment at being left alone to fend for herself and the family boiled over when she checked the refrigerator and found it stuffed with bacon and eggs and other food."

"'What do these people think I am going to do?' she demanded. 'I am *not* going to cook!'"

There was no television set for the kids to watch, no car, and if they wanted to go to the beach they'd have to cross over the highway with its fast-moving traffic. After she had complained to her husband about the unfairness of the situation he got on the phone and booked them into the Holiday Motel along with the rest of the astronauts and their families. They repacked their bags and called a cab.

"I really don't know what they expected me to do with my time there at Patrick," Betty later wrote. "I think I'd have been ready to commit suicide if I'd have stayed in that place all day waiting for him to come back home. I'd have been a complete wreck, especially with two little kids. What was I supposed to do with them? No one knew where I was. I might have gotten on the phone and said: 'Somebody come get me.' But it all just struck me wrong. I told Gus: 'This is one time I am not writing a thank-you note.' And I didn't, although I'm sure their intentions were good."[13]

"GRISSOM DID NOT BLOW THE HATCH"

When asked if he'd had much of a chance to discuss the loss of *Liberty Bell 7* with Grissom on Grand Bahama Island or afterwards, helicopter pilot Jim Lewis replied, "We saw several of the astronauts at GBI, including Gus, but other than shaking hands and passing momentary pleasantries, I didn't see Gus again until working at MSC in Houston.

"We really never discussed MR-4. I think we had moved on, and both of us knew we had followed nominal and contingency procedures properly. I had received two commendations for my actions that day and [later] Gus was subsequently selected to command the first Gemini and Apollo missions. No greater vote of assurance could have been given to him.

"We both knew he had done nothing to cause the door to detach itself that day, and we both knew we would not be able to find out specifically what happened, so there was little to discuss. Our conversations revolved around the present and future, and I'll guarantee you we had plenty to keep us occupied. The work in those days was exhilarating, intense, long, and hard … and, great fun."[14]

Flight Director Gene Kranz at his Mercury Control Center console. (Photo: NASA)

The agency's near-legendary flight controller Gene Kranz always set the tone in launch control with his calm, confident and professional manner. He would display very little emotion in the thick of a mission and he always remained focused on the tasks that lay ahead, like a general before a battle. He was not one to quibble or stay silent if he felt someone had underperformed, and he is quite adamant that Grissom did not blow the hatch.

"I spent a lot of time with Gus," Kranz stated. "Everybody alleges that the guy panicked. Gus is *not* the kind of guy who would panic … he is a very controlled person. I also knew we had an inherently different hatch design, from the standpoint of a release mechanism, to the other [Shepard] one. I knew the limitations in testing, and if Gus says he didn't do it, then he didn't do it. It's that straightforward."[15]

Another front-line exponent of Gus Grissom was McDonnell engineer Guenter Wendt, who helped insert Grissom and the other early astronauts into their spacecraft prior to hatch closure. Wendt, who died in May 2010, was a staunch admirer of Grissom. "We cannot prove what happened," he told interviewer Jim Banke in June 2000. "It was an unexplained anomaly. But we know that Grissom did not blow the hatch."

Former McDonnell Pad Leader Guenter Wendt (Photo: NASA)

Based upon his interview with Wendt, Banke later wrote that to detonate the ordnance, either Grissom would have had to firmly bang his wrist on a plunger inside the capsule, or a recovery diver alongside the spacecraft in the water could move a small panel on the outside and pull a T-shaped handle in the event that the astronaut was disabled. Later experience would show that if a Mercury astronaut were to detonate the hatch from the inside, the amount of force necessary to hit and activate the plunger would leave a nasty bruise, which Grissom didn't have.

It was put to Wendt that perhaps the switch on the outside of the capsule was accidentally pulled. Wendt responded with a theory that the small panel on the outside of *Liberty Bell 7* might have broken off as the spacecraft deployed its main parachute or shortly thereafter. In one transmission to Alan Shepard in the Mercury Control Center, Grissom said "you might make a note" of the fact that there was a six-by-six-inch hole in the parachute which Wendt said approximated the size of the access panel. Then, after splashdown, Wendt believes something may have tugged on the exposed handle just enough to cause the hatch to blow – perhaps a parachute line or a line associated with the green dye markers deployed from the capsule after splashdown. It is a known fact that after Grissom quickly egressed from the sinking spacecraft he reported becoming tangled in a marker line outside the hatch.

"That is the one [possibility] that I believe in," Wendt concluded. "It is the most logical explanation. Can we prove it? No."[16]

When asked to characterize Gus Grissom in the light of later criticism of him and his actions that day – particularly in the movie adaptation of Tom Wolfe's book, *The Right Stuff* – Jim Lewis expressed his particularly strong recollections and feelings.

"Gus flew 100 combat missions in Korea. He was a successful test pilot. He had been selected to be an astronaut. Many applied, few were chosen. He was selected to fly the second manned U.S. space mission. He was later selected to command both the first Gemini mission and the first Apollo mission. Those kinds of things do not happen to a 'screw up.'

"That kind of person would never have survived combat or been a test pilot, and would not have been selected to be the first in line to blaze the way for new space programs … Gemini and Apollo. NASA obviously had confidence in Gus. I am sad that Wolfe and his media apparently chose to ignore what, to me, is the obvious. I went through the same flight school as Gus – a bit later – flew in the Far East, and was an engineering test pilot. Nothing in Wolfe's book about flight school or the MR-4 mission or flying in general was characterized the way I would have chosen … but Wolfe neither interviewed me nor asked my opinion.

"In addition, think about this. MR-4 had a large window – the first spacecraft to have such – adjacent to the hatch. When the capsule was floating, Gus looked right out that window and could see water above the hatch sill and above the lower edge of the window, which was lined up with the lower sill of the hatch. Do you think anyone would have purposefully released a hatch under those conditions? I would add that since we had practiced such things, he also knew that I wasn't there yet and obviously hadn't lifted his spacecraft clear of the water. So then, did he accidentally hit the release? NASA records show that every astronaut who used that plunger to release a hatch got a bruise or skin abrasion from the rebound of the plunger. Gus's post-flight physical documented that his body was totally unmarked. This is positive evidence that he did not 'accidentally' hit that plunger. Had he done so, he would have been even less able to escape its rebound than any of those who actuated it on purpose.

"Gus was a consummate pilot, a very bright individual, and a skilled engineer who had everyone's respect. No one who knew him could or would argue with that statement, and that is how he should be remembered."[17]

INTERVIEWING THE ASTRONAUT

Two months before the MR-4 flight, Capt. Wayne Koons was the prime recovery helicopter pilot along with Lt. George Cox for the retrieval of Alan Shepard and his *Freedom 7* spacecraft after the MR-3 mission. At the time of the Grissom flight he was assigned to the Mercury Control Center at Cape Canaveral along with recovery manager, Robert F. ('Bob') Thompson, who needed to know what had gone wrong and caused the loss of *Liberty Bell 7*.

Robert F. Thompson (center), NASA's Recovery Coordinator with Rear Adm. W.C. Abhau (left) and Flight Director Christopher C. Kraft, Jr. (Photo: NASA)

Once Grissom was safely back on the USS *Randolph*, Robert Gilruth and Walt Williams came over to talk with Thompson, wanting to know what had gone wrong. "I don't know," Thompson admitted. "I'm not sure, but I'll go and find out and let you know." He asked Koons to accompany him on the one-hour flight out to Grand Bahama Island on a Navy administration S-2F Tracker that was kept on a skid strip at the Cape. On the way to GBI, Thompson contacted the *Randolph* and asked that helicopter pilots Lewis and Reinhard also be brought to the island for debriefing. Thompson and Koons arrived there about the same time as Grissom and the two Marine pilots.[18]

"We got down there, and they had just brought Gus in from the ship and taken him to the little Air Force medical facility," Koons would later say of that day. "He was pretty tired and uncomfortable. As I recall, he was set up to debrief on mission phases, and after the mission he was supposed to use cue cards to talk into a tape recorder and talk through pre-launch and then talk through launch, and then … through descent and landing. Bob said, 'Gus, help us out here. Would you mind doing your last card first?'"[19]

Thompson says he then asked Gus to go into a small, private room in the front of the barracks building. He discussed the events of that day with the two helicopter pilots, "got their briefing pretty quickly, and then went in and sat down and talked to Gus, just the two of us in the room. He sat on one bed and I sat on the other. So I talked to Gus about what went on. Well, after about five minutes of talking to Gus and the little bit of conversation I had with the helicopter pilots, I was pretty sure what the problem was. As far as I'm concerned, the problem was Gus got out of sequence. We had two safety devices on the door-activating mechanism. Now, this is something that we never did make a big to-do over after it was all over with, and I'm just telling you factually what went on. To open the door of the capsule after it landed, the two safety devices, you had to take a cap off … that covered the plunger that fired the door.

You had to take that cap off, turn it ninety degrees and take that off. Then you had reach up and put your finger in a ring and pull a pin out of the shaft on the plunger that fired the door, and then push a little fifty-cent-size plunger in. Once that plunger went in about an inch, it lined up with a hole that the firing pin came through that fired the door.

"So the picture here is, you've got a door-opening mechanism with two safety devices on it. The procedures were, he was supposed to stay there and not activate either one of these safety devices until the helicopter told him he had 1,800 rpm, which raised him above the water. Then he was supposed to take his helmet off, put his neck dam on, take his ECLS loose, close this, take the cap off, pull the pin, slide the plunger in, blow the door, sit on the sill and go out.

"Well, he wanted to do such a good job, that while he was waiting for *Hunt Club* to get everything ready, he says, 'I'll just take the cap off, and I'll pull the pin, but I won't push the plunger.' But now he's in a bobbing capsule with all kinds of stuff in there. Did he push the plunger? Of course not, you know, but it's kind of like you had a gun with two safeties on it. You took the two safeties off and you put it up and you pointed it at somebody, but you didn't pull the trigger, right? So he merely got out of sequence, trying to do such a good job.

"It's pretty clear to me what happened, but I agree with Gus. No, he didn't push the plunger. Did he get out of sequence? Yes. He told people that he got all ready, and he just shouldn't have done it until he was told to do it. It's just that simple. But there was no point in making a federal case, and we went on about our business. So I went to the Cape that night, found Bob Gilruth, went out in the parking lot, told him what had happened, and we went on with the program."[20]

FURTHER THOUGHTS

Samuel T. ('Sam') Beddingfield was an aeronautical engineer who was involved in testing airplanes for the U.S. Air Force at Wright-Patterson AFB before linking up with NASA at Langley Field, Virginia. At first he wanted to become involved in airplane testing once again, but the NASA interviewer suggested he might instead find better work with the rocket people, who were in need of experienced engineers. Beddingfield drove around to that area of the field and the first person he bumped into was Gus Grissom, with whom he had tested airplanes at Wright-Patterson, and the recently selected astronaut convinced him to join the rocket team. As he told interviewer Lori Walters in 2001. "I was at Langley Field, Virginia two weeks and they told me they needed me to go on a temporary trip. And so they sent me down here to Cape Canaveral to help get Project Mercury started down here and that was very early October 1959 and I've been here ever since."

Beddingfield helped establish NASA's engineering facilities at the Cape as well as administering the setting-up of Hangar S as a work area and crew quarters for the astronauts. He then became involved as a mechanical engineer in the early Redstone launches of unmanned production Mercury spacecraft, and the 5 May 1961 launch of Alan Shepard aboard *Freedom 7*.

Gus Grissom at the launch facility with spacecraft test conductor Paul Donnelly and Sam Beddingfield (right). (Photo: NASA)

Asked if he had talked over the loss of *Liberty Bell 7* with Grissom, Beddingfield said they had discussed it at length. "Yes, I talked to him quite a bit after the flight because a lot of people thought he must have fired the explosive that blew the hatch off. I knew if he had done that he'd tell me. We had tested airplanes enough together in the Air Force that when anything went wrong we knew we had to tell each other about it. And they put me on a team to go interview Gus as to what happened. And he told me he did *not* fire that hatch."

Beddingfield stated that during Mercury tests of the explosive hatch, and on the subsequent Mercury manned flights, the blowing of the hatch caused noticeable deep bone bruising on the back of the hand of the astronauts involved. Mercury astronaut Deke Slayton agreed. "No one should have the idea that Gus was going around being defensive about this hatch thing," Slayton remarked. "But he told me, sure, there was a possibility he had banged the thing by mistake …. All I know is that when Wally Schirra blew the hatch on his [MA-8] flight, he wound up with a big bruise on his hand. Gus never had one."[21]

Schirra had ridden inside his *Sigma 7* spacecraft as the recovery helicopter lifted it out of the ocean and flew it to the deck of the USS *Kearsarge* (CV-33). He blew the hatch only when he was ready to exit the capsule. He had to hit the plunger with five or six pounds of fist force; so hard that he injured his hand. He was not slow to show the tell-tale impact bruising and cut on his hand at his medical debriefing.

After deliberately blowing the explosive hatch of his *Sigma 7* spacecraft aboard the USS *Kearsarge*, Wally Schirra is assisted in his egress. (Photo: NASA)

As Schirra wrote in his autobiography, *Schirra's Space*, which was co-written with Richard Billings, "I blew the hatch on purpose, and the recoil of the plunger injured my hand – it actually caused a cut through a glove that was reinforced by metal. Gus was one of those who flew out to the ship and I showed him my hand. 'How did you cut it?' he asked. 'I blew the hatch,' I replied. Gus smiled, vindicated. It proved he hadn't blown the hatch with a hand, foot, knee or whatever, for he hadn't suffered even a minor bruise."[22]

Beddingfield concurs. "Gus did not have [a bruise] in his hand. And when we got the spacecraft back we found out that the hatch could have done something that we don't even understand. But Gus did not fire it. We were fairly comfortable in that."

Sam Beddingfield worked hard trying to determine the cause of the blown hatch. Grissom assisted by participating in extensive tests where he intentionally bumped against the plunger, but failed at all times to blow the hatch. The design engineers tried everything, but could not replicate whatever malfunction had caused the hatch to blow.

According to a Project Mercury Status Report for the period ending 31 October 1961:

During a period between August 5, 1961 and October 12, 1961 a series of environmental tests were conducted on the explosive hatch. Individual pieces of the mild detonation fuse (MDF) cord, detonator caps, and RDX lead cups were subjected to simulated altitudes of 118 miles and 135 miles and subjected to 2,000-volt +1.2 to 2.0 milliampere static discharges. No inadvertent ignition occurred.

The units were then assembled into igniter assemblies and fired by pulling the lanyard. Full-order ignition occurred. Additional MDF cord was subjected to varying exposure in hydrogen peroxide. One condition resulted in a low-order detonation without igniting the full length of 12 inches. Two repeats of the same condition failed to induce any detonation. The MDF was reduced to puddles of lead in all of these tests.

Three inert igniter assemblies were subjected to push tests with the shear pin removed, with and without vacuum, and with and without the 'O' ring. The minimum push force was 2.63 pounds. The assembly with the minimum push force was subjected to vibrations of 0.03 to 10 G at frequencies from 5 to 2,000 cycles per second with no displacement of the plunger noted. A loaded hatch assembly which was subjected to a saline solution soak, with vacuum, electrostatic shock and vibration was degraded to the point of "no fire" due to salt concentration degrading the detonator caps. This hatch assembly was then disassembled, reloaded and subjected to a simulated launch, three orbits, and reentry temperature test conditions. The pressure altitude during the test was 240,000 feet.

Upon removal from the test chamber, the hatch was subjected to a saline solution soak and repeated electrostatic discharges. No detonation occurred. The hatch was then fired by lanyard pull and normal operation occurred.[23]

In 1965 Dr. Robert Voas, who was serving as the astronauts' training officer at the time of MR-4, remained convinced that Grissom did not blow the hatch, either intentionally or accidentally. "When John Glenn completed his [MA-6] flight, he egressed from the capsule by actuating the explosive mechanism that exploded off the hatch. When he was later examined a bruise was found on his hand, caused by a pin that jumps back. On the next flight, Scott Carpenter climbed out of the top of the capsule and didn't use the hatch. Wally Schirra and Gordon Cooper both exploded the hatch and both suffered bruises on their hands. Everyone who has actuated that explosive hatch has gained a bruise. The fact that Gus did not have a bruise is final demonstration that he did not inadvertently actuate the mechanism. Although it has never been explained what did cause the accident, he has been completely absolved of the responsibility."[24]

A REVEALING POSSIBILITY

In the late 1990s, space historian Rick Boos conducted an extensive interview with Jim Lewis's co-pilot John Reinhard. On being asked how far their helicopter *Hunt Club 1* was from *Liberty Bell 7* when the hatch suddenly blew out into the water, Reinhard responded, "We were actually about five feet away and maybe not even that far. We were hovering, just sitting there waiting in a hover, that's all. Whether [Jim] considered 'final' as sliding over to [the capsule] I don't know. Had we been further away we wouldn't [have made the attempt], we would just have lost it and started the harness down and picked [Gus] up."

As to Reinhard's own version of what happened, he told Boos, "There was very little happening on final approach. We were in a hover, waiting on Gus to complete what he had to do inside the cockpit. We were ready anyway. I had what you call a 'cookie cutter' all ready to go, which was the antenna snipper. Gus said he was ready and we slid over, and when I touched the antenna there was an [electrostatic] arc. The cookie cutter had two blades on it with two explosive squibs and both of them had gone off without my activating them, and they were on two separate switches! So whatever it was that caused the arc set them both off; and fortunately the antenna was inside the yoke on it and snipped it and cut it off. At the same time that's when I was aware that Gus was coming out."

Asked if he actually saw the hatch go, Reinhard replied, "I just saw a flash and Gus came out. [The hatch] went out to the end of the lanyard and kept going. It was right after I cut the antenna." He also told Boos that Capt. Phil Upschulte (who died in April 2012) and Lt. George Cox saw it all happening from *Hunt Club 2*.

The next question was an obvious one. "So it was when you cut the antenna that the hatch blew?"

Reinhard answered without hesitation, "Right! When I touched the antenna there was an arc and both cutters fired. At that same time, the hatch came off. It could be that some static charge set it off."[25]

There is a noteworthy precedent to Reinhard's story, as told by Peter Armitage, an AVRO engineer working with NASA's Space Task Group and project manager for the Mercury capsule air drop tests. Armitage was tasked with finding a safe and manageable means of recovering Mercury space capsules from the water. As doing so from a

Although indistinct and taken from a long range, this dramatic still from footage of the recovery attempt is one of a sequence that actually shows the hatch blowing and flying away from *Liberty Bell 7* just as John Reinhard attempts to cut the HF aerial. In this still, the hatch is shown at the top of its shallow arc just above the end of the investigator's index finger. (Photo: Rick Boos, taken from NASA film footage)

Peter Armitage (left) with Gus Grissom during a spacecraft recovery exercise. (Photo: NASA)

Navy destroyer in generally rolling or choppy seas could be an extremely difficult task, helicopter recovery became the preferred option. But there was one obstacle to overcome, namely the telescoping HF antenna that extended out 22 feet above the spacecraft as a radio recovery aid.

"It started out, aluminum of about an inch or so [in diameter] and ended up just a thin antenna at the top. But when we flew in with a helicopter, [the antenna] was so high up that you couldn't get close enough because it might hit the rotor blades. And the helicopter guy had to come in with what amounts to a shepherd's crook and hook onto the spacecraft and then get the big hook in and pull it all out.

"Milt Windler and I started thinking about that problem, and we got some money out of petty cash… went down to the hardware store in Hampton and we bought two things. We bought a tree pronged pruner. They're saws on the end of a pole. And we bought just a pole tree pruner, the little thing that lops off limbs. I still have it. It's at home. I use it still.

"We did a few tests to see if we could cut through [the antenna]. Well, of course, the dynamics were terrible, because the helicopter was doing this and the spacecraft [was] waving around. It was really difficult. So Bob [Thompson] said one day,

'Why don't you go up to the naval ordnance place at Dahlgren, Virginia.' It was just north of Hampton … So I took off one day on my own, talked to the people to see if they could put an explosive charge in this tree pruner so that once you got onto it, instead of having to pull a rope, which took too much time, that you'd press a button and the antenna would [be sliced off]. They took that project and what they produced was a great thing, and then we tested that and it worked fine. The helicopter would first fly in to the antenna, put this thing [against it], press a button, lop off the antenna, and then they could go in and pick [the capsule] up.

"Interesting story there. We had sent this [cutter] to the New River Marine Base, the big Marine base [in North Carolina] where all the helicopter support came from. We sent it down there and they had played around with it. They had put a Mercury boiler-plate out with the antenna, not in the water initially but just on the land, then flown in, cut the antenna off, picked it up.

"We were using HR-2S twin-engine helicopters, big helicopters, single-blade, twin-engine. You could reach out of the front of it. In fact, we designed and built a catamaran that was nothing more than a structure with rails that stuck right out of the front of the [helicopter], with a little basket on it, something like you see guys going up telephone poles with. And here you've got this big helicopter behind you. So we had produced this, and they'd done some tests. I went down to witness a test of this new system we designed … They said, 'Would you like to do it?' and I said 'Fine.'

"So here I am, strapped on the end of this flimsy-looking thing with this gigantic helicopter behind me, and they gave me the electronic tree-pruning device in my hands, and they flew around the field and flew in to the boilerplate spacecraft with the antenna. As soon as I touched the thing to the antenna, my hands went up like that. There's a big static shock [produced by the spinning rotor]. I looked behind and there's all these Marines laughing their heads off. Of course, you needed big gloves. They were just having fun, you know. But just terrible."[26]

A CHANGE OF PROCEDURES

Despite the loss of *Liberty Bell 7*, the MR-4 mission was considered to have been a success. Eventually, rather than delay the Mercury program, a fully mechanical hatch was designed to replace the explosive version, but it was deemed to be cumbersome to operate and far too heavy, exceeding set weight constraints. The explosive hatch would remain. What *did* change was the implementation of a new set of splashdown procedures. These required the astronaut to leave the removable firing safety pin in place until after the helicopter's recovery cable had been hooked up.

The Mercury program ended after four more flights – all orbital missions – with none of the astronauts opting to blow the hatch at sea and travel to the waiting carrier aboard the recovery helicopter. John Glenn (*Friendship 7*), Wally Schirra (*Sigma 7*) and Gordon Cooper (*Faith 7*) all elected to remain in their capsule until it had been safely secured onboard ship, following which they would blow the hatch. The only exception was Scott Carpenter, whose MA-7 overshot the planned landing zone by 250 nautical miles, which meant it would be a considerable time before he and his spacecraft could be retrieved from the sea. Rather than remain in the unstable craft with the near-certainty of becoming seasick, or blow the side hatch and risk having *Aurora 7* fill with seawater and sink, he squeezed up and out through the alternative recovery compartment exit at the top of the spacecraft.

The fact that his spacecraft had been lost at sea remained a source of irritation and torment for Gus Grissom.

"We had worked so hard and had overcome so much to get *Liberty Bell* launched that it just seemed tragic that a glitch robbed us of the capsule and its instruments at the very last minute. It was especially hard for me, as a professional pilot. In all my years of flying – including combat in Korea – this was the first time that my aircraft and I had not come back together. In my entire career as a pilot, *Liberty Bell* was the first thing I had ever lost. We tried for weeks afterwards to find out *what* happened, and *how* it happened. I even crawled into capsules and tried to duplicate all of my movements in order to see if I could make it happen again. But it was impossible. The plunger that detonates the bolts is so far out of the way that I would've had to reach for it on purpose to hit it – and this I did not do. Even when I thrashed about with my elbows, I couldn't bump against it accidentally. It remained a mystery how that hatch blew. And I am afraid it always will. It was just one of those things."[27]

Four days after the MR-4 mission, a McDonnell engineer occupies a test capsule showing an astronaut's position in relation to the explosive hatch detonator, seen above his head. (Photo courtesy the Kansas Cosmosphere and Space Museum)

ON TO GEMINI

As of 15 July 1962, Grissom was promoted to the rank of major. Four days later he proudly received the Gen. Thomas D. White Trophy as the "Air Force member who has made the most outstanding contribution to the nation's progress in aerospace." Also in 1962, NASA finally moved its manned space flight operations from Langley to the newly built Manned Spacecraft Center (MSC) in Houston. The Grissoms were now able to build their first real home, a three-bedroom house in the Timber Cove development near Seabrook.

Following his MR-4 flight, Grissom acted as CapCom for later Mercury flights, but spent a lot of his time in St. Louis assisting McDonnell in the development and construction of the Gemini spacecraft. In October 1962 Deke Slayton also assigned Grissom to supervise the nine Group 2 astronauts in the lead-up to Project Gemini. A practical man, Grissom had understood early on that a second Mercury mission was not on the cards for him, as the remaining astronauts would each be assigned flights that would complete the program. "By then Gemini was in the works," he wrote in his memoir, "and I realized that if I were going to fly in space again, this was my opportunity, so I sort of drifted unobtrusively into taking more and more part in Gemini."[28]

Grissom then turned his efforts to the Gemini spacecraft and specifically the layout of the cockpit controls and instrumentation. He was instrumental in having everything laid out as a pilot would like; so much so that the spacecraft soon became known as the

Gus Grissom is congratulated by Air Force Secretary Eugene M. Zuckert on being presented with the Thomas D. White trophy. (Photo: UPI)

Alan Shepard and Gus Grissom autograph an improvised space helmet made by a New York Boy Scout. (Photo: NASA)

"Gusmobile." However, later in the program taller astronauts such as Ed White and Tom Stafford complained that although the arrangement might have been ideal for Grissom's compact five-foot seven-inch frame, it was something of a tight squeeze for them.

As recalled by Betty Grissom, "Gus was living with the Gemini spacecraft, putting the flier's touch on the vehicle in which men might live and maneuver in space for days or weeks. At the McDonnell plant in St. Louis he sat in the mockup for hours, day after day, testing each switch and control knob, applying the aviator's instinct which would guide engineers and technicians in making the machine more flyable and habitable. Although his advice usually was quiet and brief, the engineers never doubted that he meant business, and never forgot that the inanimate mass of metal, wiring and black boxes was a flying machine to carry men from Earth and back again. In fact, they were laying bets that Gus would be the first to fly it."[29]

Then, in April 1964 Grissom's diligent work was rewarded when he received his second mission assignment as commander of the first Gemini two-man mission. The official announcement came on 13 April, five days after the successful launch of the unmanned Gemini 1 capsule. He was the first astronaut to be officially assigned to a second space flight.

Among many other honors, Grissom received a presentation trophy from the U.S. Junior Chamber of Commerce as "one of the ten outstanding young men of 1961." (Photo: NASA)

Wally Schirra, selected as the *Gemini-Titan 3* (GT-3) backup commander along with co-pilot Stafford, was delighted that his Mercury colleague and friend had been given the first flight assignment, saying that he felt Grissom now had something to prove. "He was angry about being blamed for his spacecraft having sunk, and he was fighting to come back out of the pack. Gus was a tiger. He wanted the first Gemini flight, and by God he got it."[30]

Grissom shows his trophy to NASA Operations Director Walt Williams. (Photo: NASA)

GUS AND THE DOCTORS

Gus Grissom admired dedication in most people, but there was one particular profession that caused him to be wary in their presence. Normally outspoken and gregarious people, NASA's astronauts would readily discuss most things in order to get the answers they sought – except when it came to the space agency's medical staff. They knew from their days of flying high-performance aircraft that any hint of a complaint relating to their wellbeing could attract the unwanted attention of the doctors.

Prior to his unexpected but ultimately lengthy assignment as a NASA flight surgeon in 1960, Robert H. Moser was an Army major taking a fellowship in hematology at the Utah Medical School in Salt Lake City. As he recalls, little was known at that

In October 1963, Grissom was the proud recipient of an award from the Air Force Communications Service in recognition of becoming the first Air Force officer to receive astronaut wings for his flight into space and communicating with ground stations. (Photo: NASA)

time about a human being's psychological and physiological capability to endure and function in space. He and his fellow Army officers would get to know the Mercury astronauts quite well over the years, but there was always something of an unspoken chasm between an astronaut and any medical practitioner.

In September 1962 Dr Moser was stationed on the island of Kauai in preparation for the six-orbit Mercury flight of Wally Schirra the following month. Occasionally, he and Gus Grissom would spend time kicking back together in a local bar.

"We had become chums during the seemingly endless simulated missions that always preceded orbital flights," Moser once reflected for *The Pharos* magazine. "This was a rare downtime. Gus was our CapCom and I was the medical flight controller. I asked him why flyers hated doctors. He straightened himself on the bar stool, and peered into the bottom of his glass. 'I'll tell you, doc. When you walk into the flight surgeon's office, you *have* your ticket. When you walk out, you might *not*.'"

Moser pressed further. "I asked him, 'Gus, if you were sitting on top of that big firecracker and the countdown got to about minus seven, and suddenly you felt the worst sort of pain imaginable pressing down on your mid-chest and radiating down the inside of your left arm, what would you do? Would you let us know?'

"Gus took a long moment to gaze at the bay. 'Only if I thought I was going to die.'"[31]

Gus Grissom with an early model of the Gemini spacecraft. (Photo: NASA)

References

1. *The Milwaukee Sentinel* newspaper, Wisconsin, "Astronaut Whoops it Up," issue 22 July 1961, p. 3
2. *Spokane Daily Chronicle* (Washington) newspaper, "Fear is Dominant Emotion in Grissom's Indiana Home," issue 21 July 1961, p. 11
3. *Spokane Daily Chronicle* (Washington) newspaper, "Spaceman's Brother Now Able to Relax," issue 21 July 1961, p. 11
4. *Spokane Daily Chronicle* (Washington) newspaper, "Wife Confident of Success," issue 21 July 1961, p. 11
5. *Warsaw Times-Union* (Indiana) newspaper, "Gus Grissom Day in Indiana," 21 July 1961, p. 2

6. *Spokane Daily Chronicle* newspaper (Washington), "Space Bill is Approved by Kennedy," issue 21 July 1961, p. 11

7. Indiana Historical Society online article, "Virgil (Gus) Grissom" by HIS Staff, website: http://www.indianahistory.org/our-collections/library-and-archives/notable-hoosiers/virgil-gus-grissom]

8. E-mail correspondence Dee O'Hara with Colin Burgess, 16 July 2013

9. *The Spartanburg Herald* newspaper (South Carolina), article, "Rest, Checkups, Reports Now Lie Ahead For Second U.S. Spaceman," issue 22 July 1961, p. 3

10. E-mail correspondence Rick Boos with Colin Burgess, supplying interview by him with John Reinhard, circa 1998

11. *The Milwaukee Sentinel* newspaper (Wisconsin), article, "Pow, Water Rushed In: Gus," issue 23 July, p. 2

12. Grissom, Betty and Henry Still, *Starfall*, Thomas Y. Crowell Company, New York, NY, 1974

13. *Ibid*

14. E-mail correspondence James D. Lewis with Colin Burgess, November 2002 to August 2003

15. Kranz, Gene, interview conducted by Francis French, San Diego, California, 22-23 March 2001

16. Guenter Wendt interviewed by Jim Banke for *Space.com* article, "Gus Grissom Didn't Sink Liberty Bell Seven," 17 June 2000

17. E-mail correspondence James D. Lewis with Colin Burgess, November 2002 to August 2003

18. Thompson, Robert F., interviewed by Kevin M. Rusnak for NASA JSC Oral History program, Houston, Texas, 29 August 2000

19. Koons, Wayne, interviewed by Rebecca Wright for NASA JSC Oral History program, Houston, Texas, 14 October 2004

20. Thompson, Robert F., interviewed by Kevin M. Rusnak for NASA JSC Oral History program, Houston, Texas, 29 August 2000

21. Beddingfield, Samuel T., interviewed by Dr. Lori C. Walters for the U.S. Space Walk of Fame Foundation Oral History Program, Titusville, Florida, 26 October 2001

22. Schirra, Walter M, Jr., with Richard N. Billings, *Schirra's Space*, Quinlan Press, Boston, MA, 1988

23. NASA Project Mercury Status Report No. 12 for Period Ending October 31, 1961

24. Voas, R. Robert, interviewed by Shirley Thomas for *Men of Space* (Volume 7), chapter "Virgil I. Grissom," Chilton Company, Philadelphia, PA, 1965, p. 105

25. E-mail correspondence Rick Boos with Colin Burgess, supplying interview by him with John Reinhard, circa 1998

26. Armitage, Peter J., interviewed by Kevin M. Rusnak for NASA JSC Oral History program, Houston, Texas, 20 August 2001

27. Carpenter, Malcolm S., Cooper, L. Gordon, Jr., Glenn, John H. Jr., Grissom, Virgil I., Schirra, Walter M., Jr., Shepard, Alan B. Jr., and Slayton, Donald K., *We Seven*, Simon and Schuster Inc., New York, NY, 1962, p. 298

28. Grissom, Virgil, *Gemini! A Personal Account of Man's venture Into Space*, The Macmillan Company, Toronto, Ontario, 1968, p. 73
29. Grissom, Betty and Henry Still, *Starfall*, Thomas Y. Crowell Company, New York, NY, 1974
30. Schirra, Walter M, Jr., with Richard N. Billings, *Schirra's Space*, Quinlan Press, Boston, MA, 1988
31. E-mail correspondence with Robert H. Moser, 16 July 2009, and permission to quote from Moser, Robert H., article "My Romance with Space," in *The Pharos* (of Alpha Omega Alpha Honor Medical Society) magazine, Autumn 2003, pp. 11-17.

7

A tale of two hatches

On 7 December 1961 Robert Gilruth, director of the Manned Spacecraft Center in Houston, announced plans for a spacecraft that would be piloted by two astronauts and would advance the United States to the next level of manned space flight. This new program would help to develop manned space flight rendezvous techniques in a more spacious craft capable of docking with other vehicles while in Earth orbit. This "Advanced Mercury" concept would also serve as a bridge between the Mercury and Apollo lunar landing programs. Since no project title had been officially assigned to the new program, it was simply referred to as Mercury Mark II.

A NASA bulletin reported that the agency would negotiate with the McDonnell Aircraft Corp. of St. Louis as prime contractor for the new spacecraft. Weighing about two tons – twice that of the Mercury capsule – the spacecraft was intended to be launched atop a new booster, the Air Force Titan II, constructed by the Martin-Marietta Company. The rendezvous target to be used during the program was to be an Agena stage produced by the Lockheed Aircraft Corporation, and this would be launched by an Atlas rocket. Preliminary cost estimates for the program, including about a dozen spacecraft, Atlas-Agena and Titan II vehicles, was in the vicinity of $500 million. As the NASA bulletin suggested:

> Two-man flights should begin in 1963-64, starting with several unmanned ballistic flights from Cape Canaveral for tests of overall booster-spacecraft compatibility and systems engineering. Several manned orbital flights will follow. Rendezvous flybys and actual docking missions will be attempted in final phases of the program.
>
> This program provides the earliest means of experimenting with manned rendezvous techniques. At the same time, the two-man craft will be capable of Earth-orbiting flights of a week or more, thereby providing pilot training for future, long-duration circular and lunar landing flights.
>
> NASA's current seven astronauts will serve as pilots for this program. Additional crew members may be phased in during later stages.[1]

THE FIRST GEMINI CREW

The following month the new program had been given a name. On 3 January 1962 NASA announced the two-man spacecraft would be called Gemini, the Latin word for "twins." This followed a suggestion by Alex P. Nagy from NASA's Office of Manned Space Flight at the agency's Washington Headquarters, who not only had the distinction of naming the nation's new space program but also of receiving the associated prize of a bottle of scotch whiskey. Appropriately enough, Gemini was the name given to the third constellation of the zodiac (the sign in astrology that is controlled by Mercury) and comprised of two stars called Castor and Pollux. The Gemini spacecraft

Dr. Robert Gilruth of NASA's Manned Spacecraft Center in Houston. (Photo: NASA)

In preparation for future Apollo lunar missions, all eligible astronauts including Gus Grissom underwent geology training, and he is shown here in the Grand Canyon in 1964. (Photo: NASA)

Part of the fun of astronaut geology training in the Grand Canyon was riding out on mules. (Photo: NASA)

was the same high-drag shape as the Mercury capsule, but with around 50 percent greater interior room.

The first test flight (GLV-1) of a Gemini spacecraft atop a Titan II took place on 8 April 1964, using Launch Complex 19 at Cape Canaveral, and it was a complete success. The second stage of the Titan II and the attached, uninhabited spacecraft orbited the Earth 64 times, although the official part of the mission ended after only three orbits. As there were no plans to retrieve the spacecraft, the entire assembly of the spacecraft and the upper stage of the booster reentered the atmosphere four days later and burned up over the South Atlantic. All the major mission objectives had been met, principally that of testing the structural integrity of the spacecraft and the modified Titan II booster and proving that the spacecraft was capable of carrying a crew.

Liftoff of the first Gemini-Titan II flight (GLV-1) on 8 April 1964. (Photo: NASA)

As history records, Alan Shepard was actually the original choice to command the first Gemini orbital test flight with co-pilot Tom Stafford, but to his consternation he fell victim to a debilitating inner ear ailment (later diagnosed as Ménière's disease) which caused him to be medically disqualified from flying in October 1963.

Early in 1964 *Missiles and Rockets* magazine made a surprise claim. "There are unconfirmed reports that the first Gemini astronaut team will be made up of Virgil ('Gus') Grissom and Neal [sic] Armstrong. Grissom is the member of the original Mercury astronaut team who has worked most closely with McDonnell Aircraft Corp. in design and development of the spacecraft. Armstrong has chalked up many flight hours during the X-15 program. Official announcement of the first pilot team is expected around May 1."[2]

Grissom had been penciled in to command the fourth manned Gemini flight and the grounding of Shepard caused Deke Slayton (who had himself been grounded before he could make a Mercury flight and, as the newly named deputy director of Flight Crew Operations, was in charge of crew assignments) to promote Gus to the first manned mission, the three-orbit test flight designated Gemini 3. Now a suitable co-pilot was needed to partner him, and Slayton wanted to give flight experience to the nine newly selected astronauts. He subsequently paired them in the first Gemini flights with an experienced astronaut from the Mercury program.

At first the Gemini 3 mission was scheduled for launch in December 1964. Air Force Capt. Frank Borman had recently finished his astronaut training and, like the other eight pilots of the second astronaut group, was wondering when he might be assigned to a Gemini mission and which one it would be. Everyone knew that the Mercury astronauts who would fly as mission commanders on Gemini would have the power to veto any decision that Slayton might make regarding their co-pilot in order to avoid any potential clashes of personality.

It came as an unexpected but welcome surprise when Borman received a phone call one day from Grissom, who told him (although it was yet to be made official) that he, Grissom, had been named by Slayton to command the first manned Gemini mission and Borman had been tentatively assigned as his co-pilot. Grissom wanted to talk over the mission and its requirements before the final crewing decision was made. The two men arranged to meet at his house, and Borman could not get there soon enough. They spent an hour or so deep in conversation and not long after that Borman was informed of a change of crewing that meant Grissom would fly with another member of the Group 2 astronauts.

"I haven't the slightest idea what went wrong," Borman later pointed out in his autobiography, "but he apparently wasn't too impressed with me. The next thing I knew, I had been replaced by John Young, who didn't try very hard to conceal his delight, for which I couldn't blame him."[3]

To soften Borman's disappointment at the news, Slayton told him that he would instead be assigned as backup commander of the second mission, with co-pilot Jim Lovell.

USAF Capt. Frank Borman, Group 2 astronaut. (Photo: NASA)

EJECTION SEATS AND A FAULTY HATCH

Prior to the first manned Gemini mission, plans were accelerated for the suborbital flight of Gemini 2, which was to test the spacecraft's heat shield and splash into the South Atlantic carrying two instrument boxes substituting for astronauts. Unlike the GLV-1 mission, this one was jinxed. In addition to a succession delays for technical reasons, it fell prey to severe weather patterns over the Cape. On 20 August 1964 a violent electric storm hit the area and a bolt of lightning struck the Titan II, which had been erected in its gantry the previous month. Many of the delicate instruments were damaged, and the vehicle had to be dismantled in order to check thousands of vital components and revalidate its systems. A second launch attempt was aborted when the rocket had to be removed from the pad again to protect it from the fierce winds of the rapidly developing Hurricane Cleo. Soon after the rocket had returned to the pad, the prospect of a battering by Hurricane Dora caused it to be removed a third time on 11 September. And then on 9 December the Titan II suffered a launch pad shutdown due to a malfunction in the booster's hydraulic system. The mission was rescheduled for 19 January 1965. This time the launch was successful, and the spacecraft splashed down 25 miles from the recovery force. The spacecraft and heat shield were found to have performed as required, and NASA gave the go-ahead for the first manned launch.

Unlike the cramped Mercury, the Gemini spacecraft was fitted with powerful ejection seats that either astronaut could activate to escape a launch pad explosion. The seats ejected with such a tremendous thrust that the astronauts hoped never to find themselves in situation where they would have to eject. Tests were carried out using mannequins, and one of the final tests occurred on 16 January with Grissom and Young looking on as ground controllers sent the firing signal to the spacecraft. In milliseconds, powerful solid rocket motors hurled the dummy astronauts along with their seats and parachutes through the simultaneously opened hatch exits. At least, that was the plan. Both Grissom and Young winced visibly when one of the hatches failed to open and the seated mannequin was propelled straight through it. "That would give you one hell of a headache," the laconic Young later observed, "but a short one."

In the event of a launch pad explosion, the Gemini spacecraft was equipped with ejection seats to blast the astronauts 800 feet from the pad. To test the ejection and parachute systems, boilerplate capsules were mounted in launch attitude on top of a high tower and the seats carrying mannequin astronauts were fired across the desert. (Photo: NASA)

The test served only to reinforce in the astronauts their mistrust of the system, and gave Grissom another reason to dislike the word "hatch." Although a further test two weeks later worked perfectly, somewhat restoring their faith in the escape apparatus, they still hoped they would never have to use it.

THE "UNSINKABLE" MOLLY BROWN

During the Mercury program, in the same way in which service pilots personalized a particular aircraft by giving it a nickname, the astronauts had been permitted to name their capsule and use the name as the official call sign for the mission. Much to their annoyance this custom was rescinded by NASA during the Gemini program, when mission managers decided to utilize the mission name as the official spacecraft call sign.

An obstinate Grissom decided to give his Gemini spacecraft an unofficial name. At first, he and John Young toyed with the idea of an American Indian name, but then he read that the musical *The Unsinkable Molly Brown* was nearing the end of its Broadway run, and this gave him the idea for a humorous repost to the splashdown drama on his previous flight.

"I'd been accused of being more than a little sensitive about the loss of my *Liberty Bell 7*, and it struck me that the best way to squelch this idea was to kid [about] it. And from what I knew about our Gemini spacecraft, I felt certain it would indeed be unsinkable. So John and I agreed that we'd christen our baby *Molly Brown*."

John Young (left) and Gus Grissom inspect a training mockup of the Gemini spacecraft. (Photo: NASA)

While some sympathetic NASA officials found the name quite amusing, others certainly didn't, and he was told to think of something more respectable. When he responded with "Sure … how about the *Titanic*?" it became quite evident that this determined astronaut was not going to yield without a fight. A sense of resignation finally set in, and the officials partially relented. The spacecraft to be flown on the Gemini-Titan 3 mission would thereafter be known – unofficially – as the *Molly Brown*.

But this would be the last time a Gemini crew was given any sort of latitude to christen a spacecraft, although they were allowed to design personalized shoulder patches. As well, all succeeding Gemini flights would be identified using Roman numerals, starting with Gemini-Titan IV (GT-IV).

TWO MEN IN SPACE

Over several months, Grissom and Young practically lived with their spacecraft at the McDonnell plant, spending scores of hours training in simulators, memorizing every switch, knob, light, dial, and handle until they could quickly and instinctively find each one in moments.

As the first astronaut to fly both Mercury and Gemini spacecraft, Grissom offered a comparison between the two vehicles.

"The most important difference in the Gemini spacecraft is the amount of control the pilot exercises over all the functions. Gemini is the first true pilot's spacecraft. Although Mercury was handled in flight a good deal by manual control, it was designed essentially as a fully automatic machine with manual control capability as a backup to the automatic systems. We have proven that we can contribute a great deal to the successful accomplishment of the mission by controlling the spacecraft, but we had to override the automatic system to make our point.

"The original concept in manned Mercury flights was that the pilot would go along as an observer since his capabilities in space were unknown at that time. Gemini, on the other hand, demands pilot response in all its functions. The pilot must decide whether to abort a mission during the boost phase, he separates the spacecraft from the booster, he steers the craft from one orbit to another, he [does the] rendezvous with the Agena capsule, he must decide when and where to reenter the Earth's atmosphere, he must control the reentry, and then guide the spacecraft to a safe landing at a prede-termined point. Gemini will be a pilot-controlled operational spacecraft, not just a research and development vehicle.

"The escape system on the Gemini is a rocket ejection seat similar to that used on high-performance aircraft. Either the command pilot or the pilot can eject both seats simultaneously by pulling a lanyard located between his legs. There is no automatic sensing device to eject the seats. The pilot has to make his own decision to either eject and abort the mission or ride it out."[4]

Gemini 3 (GT-3) was always planned as a shakedown trial of the new spacecraft, and although it was only ever designated a three-orbit flight, GT-3 was nevertheless a crucial mission chock-full of tests and other activities right through to splashdown.

Its prime objectives were to demonstrate manned orbital flight in the spacecraft; to demonstrate and evaluate the capability to maneuver the spacecraft; to demonstrate and evaluate the operation of the worldwide tracking network; to evaluate the performance of onboard systems; and to recover the spacecraft and evaluate the recovery system.

At 9:24 a.m. (EST) on 23 March 1965, Grissom and Young were launched into orbit. According to the transcript, on liftoff CapCom Gordon Cooper even gave the spacecraft's name his own blessing by saying, "You're on your way, *Molly Brown*." To which Grissom responded, "Yeah, man ... oh, man!" At launch, Grissom had both gloved hands gripping the D-ring, ready to trigger the ejection seats at any time during the first fifty seconds of the flight, after which they were not a viable means of escape. Young, being more trusting of his commander and the spacecraft, kept his hands firmly in his lap as Grissom later reported.

The two astronauts then successfully completed a near-perfect three-orbit mission in a little less than five hours.

For Grissom, there was only one significant unanticipated incident during the splashdown. This was due to the fact that the Gemini spacecraft had been designed to land at an angle in the water rather than base-first like Mercury. Accordingly the parachute harness was rigged so that as the main parachute filled, the capsule was snapped from the vertical to a 45-degree landing attitude. Neither astronaut was prepared for the shock and severity of this action when the nose suddenly dropped after the main chute opened, and they were thrown forward. Grissom's helmet hit a knob on the instrument panel, both cracking his faceplate and making a small hole. Young's faceplate was similarly scratched following the jarring movement. Grissom later recommended that a small warning buzzer be installed to alert the crew when this action was about to take place.

The launch of the GT-3 mission carries Gus Grissom and John Young into orbit. In the process, Grissom entered the history books as the first person to be launched into space a second time. (Photo: NASA)

As a Navy helicopter hovers overhead, divers attach the flotation collar to the Gemini space-craft. (Photo: NASA)

As Grissom later recalled, when *Molly Brown* splashed down in the Atlantic, "In all honesty I must state that my first thought as we hit the water was, 'Oh my God, here we go again!' The Gemini spacecraft is designed so that the left window, my window, will be above water after landing, but instead of looking up at blue sky, I was peering down at blue water. I realized that I hadn't cut loose our parachute, and the wind was blowing it across the water, dragging us along underneath like a submarine. Remembering that prematurely blown hatch on my *Liberty Bell 7*, it took all the nerve I could muster to reach out and trigger the parachute-release mechanism. But with the parachute gone, we bobbed to the surface like a cork in the position we were supposed to take."[5]

Shortly thereafter an Air Rescue Service C-54 Skymaster deployed a pararescue team into the water, followed by another dive team dropped from a Navy helicopter, which attached the spacecraft's floatation collar. Meanwhile, as Grissom later noted, he and Young were experiencing waves of nausea in their swaying spacecraft.

"It was, to put it bluntly, hot as hell inside the spacecraft, and that, coupled with the pitching and rolling, gave both of us some uncomfortable minutes of seasickness. John managed to hang on to his meal, but I lost mine in short order. Then we climbed out of our space suits."[6]

The information that they had landed a little short of the projected splashdown site was communicated to the astronauts, along with the fact that USS *Intrepid*, the recovery carrier, was 60 miles downrange and wouldn't arrive in the area for about ninety minutes, so they decided to request a helicopter pickup rather than remain in their sweltering, swaying spacecraft.

"I left the spacecraft first," Grissom explained, "because my hatch was the one fully out of the water and could be opened without danger of flooding the cabin. John Young told me that this was the first time he'd ever seen a captain leave his ship first, so I promoted him to captain on the spot, which, he later claimed, entitled him, as a navy man, to rechristen our spacecraft the USS *Molly Brown*."[7]

After being taken aboard the recovery helicopter they were flown to the *Intrepid*, where they underwent the mandatory debriefing soon after touching down.

Grissom later said that if NASA had asked him and Young to go back into space aboard the *Molly Brown* the next day, they would have done so with pleasure.

"She flew like a queen, did our unsinkable *Molly*," he stated with a smile.

Two years prior to his Gemini mission, on 25 January 1963 Grissom gave a talk entitled "Green on Gemini" at the U.S. Air Force Academy in Colorado Springs, in which he offered encouragement to those who might seek to join the space program as future astronauts.

"The training is tough and a lot of knowledge has to be crammed into our skull," he stated. "At the same time, our bodies will be learning totally new responses, but the end result will give man a new freedom in space. Until now, man has been a

The crewless *Molly Brown* spacecraft is hoisted aboard the USS *Intrepid*, 23 March 1965. (Photo: NASA)

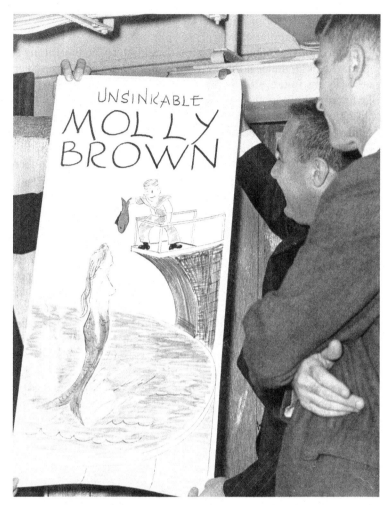

Gus Grissom and John Young admire a poster presented to them after their space flight by the Navy crew aboard the USS *Intrepid*. (Photo: NASA)

self-experimenting guinea pig, subjecting himself to space to test whether he can stand up to this hostile, new environment. With the Gemini program, man has stepped into his proper role – the explorer of space."[8]

APOLLO COMMANDER

By the end of 1965, NASA's focus was increasingly turning its attention from the highly successful Gemini series of missions to the forthcoming Apollo program. With the first manned orbital flight scheduled before the end of the following year Deke

Slayton, the agency's Director of Flight Crew Operations, decided it was time to provisionally select the first Apollo crews.

Slayton's initial choice to command the maiden flight, an Earth-orbiting test of the Apollo spacecraft would, under normal circumstances, have been Alan Shepard. But with Shepard's flying future still in doubt owing to his affliction with Ménière's Disease, Slayton opted for another of his experienced Mercury group. "Gus Grissom was going to be coming off the backup assignment to GT-6A," he explained in his memoirs, "and so was a pretty natural choice for commander of the first mission."[9] Both Grissom and Young served as backups to the prime crew of Wally Schirra and Tom Stafford for this flight, which launched successfully on 15 December 1965 to perform the first orbital rendezvous mission of the program.

Also working in Grissom's favor as command pilot for the first Apollo mission was the fact that he served as director of the Gemini program in the astronaut office before being named command pilot for the first two-man space flight. In February 1966 he was named to head the Apollo program in that office, so it seemed history would repeat itself.[10]

America's first spacewalker Ed White was another name Slayton had penciled in for the first Apollo crew. With the lunar module still under development, there was no requirement for a lunar module pilot on this Earth-orbiting test flight, so Slayton decided he could assign a 'rookie' crewman to occupy the third seat, and his choice came down to two suitable candidates – Donn Eisele and Roger Chaffee. Both were Group 3 astronauts who had earlier been paired on tests of the lunar spacesuit's life-support systems. But as the choice came down to a question of crew compatibility, Slayton decided that Eisele might be a better fit. He now had his first Apollo crew, although the three names still had to be submitted to NASA headquarters for the agency's approval and official confirmation. His judgment in selecting crews had proven rock-solid in the past, so he envisaged little or no problems in having these three men approved.

NASA astronaut Donn F. Eisele. (Photo: NASA)

Unexpectedly, fate then brought about a last-minute crew change. In September 1964 Eisele had been participating in zero-gravity training aboard a NASA KC-135 aircraft when he accidentally dislocated his left shoulder. The injury had healed, but much to his chagrin Eisele dislocated the shoulder a second time in January 1966 while taking part in some strenuous physical exercises. This fresh injury prompted Slayton to replace Eisele with Chaffee on the crew list that he submitted to NASA headquarters.

Disappointed, but determined to make good and be reassigned, Eisele eventually overcame his shoulder injury and was provisionally assigned to the second planned Apollo mission along with Wally Schirra and Walt Cunningham. In view of some conjecture regarding early Apollo crewing, these crew changes were all verified by Eisele's first wife, Harriett, by Walt Cunningham, and – prior to his death in 2007 – by Wally Schirra.

For Gus Grissom, Apollo 204 (as it was then designated because it would be the fourth launch of the Apollo IB rocket; 'Apollo 1' would be applied later) provided him with the chance to command a second test flight after the unqualified success of Gemini 3. It would also offer further vindication of his character and courage after all the rumors he had endured that he had panicked and blown the hatch on *Liberty Bell 7* five years earlier. Had that been the case, it is highly unlikely he would have been awarded the first flight in the Gemini series, let alone the maiden test flight of the Apollo spacecraft whose duration was open-ended up to around two weeks.

In November 1966, Grissom penned a widely syndicated column concerning the upcoming mission in which he revealed his hopes for the mission and for America's next steps in space. Reproduced here in part, it was published under the title *Three Times a Command Pilot*.

Robert Gilruth (far right) introduces the crew of the first Apollo mission, Roger Chaffee, Ed White and Gus Grissom. (Photo: NASA)

In *Liberty Bell 7*, I was a man in a can just along for the ride. *Molly Brown*, bless her heart, was a machine I could maneuver. And now in Apollo 204, Ed White, Roger Chaffee and I will be in a spacecraft designed to go to the Moon and back.

Soon I'll be the first United States astronaut to make three flights – one in each of our first three space programs. My upcoming flight is in an Apollo spacecraft which makes my old Mercury *Liberty Bell 7* look something like an early flivver [cheap automobile]. But in those days we weren't all that concerned about maneuverability. We were out to discover whether man could survive G-forces of liftoff and the environment of space. And we learned that man could survive. During the past two years the Gemini program has taught us that we can fly our spacecraft, rendezvous and dock, and even perform meaningful tasks outside the spacecraft.

My fellow crewmembers and I are finding that our Apollo spacecraft is infinitely more complex than Gemini or Mercury. And so is the flight plan, even for our own Earth orbiting mission.

Our job will be to operate and observe and evaluate all of the spacecraft systems. When necessary, we must come up with suggestions for solutions to any problem we encounter. And this we can only do in actual space flight. We may spend anywhere from three to fourteen days in orbit, possibly longer, learning as much as we can about the spacecraft's performance. Even as we fly the mission, people on the ground will be working to make the Apollo lunar spacecraft an even more sophisticated vehicle than ours.[11]

A FATAL FIRE

By late January 1967, despite a variety of frustrating problems with their spacecraft and its systems, the three astronauts were fully trained and ready to fly on the Earth-orbiting test flight of a spacecraft that, with modifications, would one day carry three American astronauts to the Moon. As professional test pilots they were aware of the dangers they faced. During a 1966 address to the Associated Press, Grissom openly discussed his feelings on those dangers. "If we die," he stressed, "we want people to accept it. We are in a risky business and we hope that if anything happens to us, it will not delay the program. The conquest of space is worth the risk of life."[12]

Grissom's words would prove sadly prophetic, and came tragically true during a mated test of the Apollo spacecraft and Saturn IB rocket on 27 January.

Three critical objectives had to be met before the scheduled launch of Apollo 204 on 21 February 1967. They were a "Plugs Out" test, the Flight Readiness Test, and the Countdown Demonstration Test. The test on Friday, 27 January was the "Plugs Out". Although the spacecraft cabin would be internally pressurized by 100 percent oxygen, it was not considered a particularly dangerous test of the spacecraft and its systems because the inert Saturn IB rocket was not loaded with fuel. A fuelled test would occur only as the final "wet" mock test immediately preceding the February launch.

On the morning of 27 January, technicians at KSC and MSC began the well-rehearsed task of checking the spacecraft systems for the test. By this time the spacecraft had undergone 20 weeks of tests and checkout at the Downey plant in California, plus an additional 21 weeks of checks and modifications at the Cape.

At 7:42 a.m. on that fateful day, technicians began powering up the spacecraft, sending electric current surging through nearly 30 miles of wiring coiled in thick bundles around the floor of the spacecraft and through enclosed recesses above and below the three contoured couches the astronauts would occupy. After lunch, once all was in readiness, the crew was driven by van to the launch pad and made their way up to the spacecraft level on the massive gantry, ready for a long afternoon of checks and tests. Once inserted into their respective couches they plugged into the spacecraft's communications and oxygen systems. It was now 1:19 p.m. The pad technicians then sealed the capsule's pressure vessel inner hatch which, unlike the outward-opening hatches used during Mercury and Gemini, opened inwards above Ed White's head. Once it had been secured, the hatch was held in place and sealed by a series of clamps. Next the technicians secured the much heavier, cumbersome outer crew access hatch. To complete the process, they locked the fiberglass-and-cork booster protection cap in place.

To exit the spacecraft, extensive ratcheting of the inner hatch by a torque wrench was required in order to retract six dog-leg locking bars. At the same time, the cabin pressure had to be reduced by operating a purge valve, which would then allow the hatch to be opened inwards, in much the same manner as a modern airliner. Once the hatch had been hauled into the cabin, White could then operate a quick-release mechanism that unlocked the outer hatch.

Initiating the day's test, the crew, now strapped and plugged in, began to purge their spacesuits and the spacecraft of all gases except oxygen. The cabin was to be

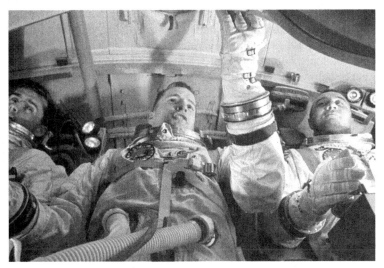

During a training session Chaffee, White and Grissom run through ground checks aboard Spacecraft 012. (Photo: NASA)

pressurized above ambient in order to simulate normal flight conditions and ensure no contamination from outside. This involved raising the cabin pressure to 16.7 psi (pounds per square inch) of pure oxygen.

As the astronauts worked their way through a series of checklists, their irritation – particularly in the case of Grissom – grew as minor glitches disrupted the check-out sequence. He then reported a foul odor in the space suit loop, which he described as "a sour smell somewhat like buttermilk." Adding to their displeasure, the crew had problems trying to communicate with the control center, which extended to include irregular communications between the Operations and Checkout Building and the blockhouse at Pad 34. Becoming increasingly agitated, Grissom reached the point where he vented his frustration. "How the hell can you expect us to get to the Moon if you people can't hook us up with a ground station?" he growled at one point. "Get with it out there!"

At 6:20 p.m., during another in a series of holds, there is evidence to suggest that Grissom decided to do something about the communications problem without telling the control center. He unbuckled his seat harness and disconnected his cobra cable in an attempt to check connections below his feet. The cobra cable was a multi-wired communication cable within a sheath which connected each of the three astronauts to the instrument panel. Further evidence indicates that he then eased himself down into the cramped lower equipment bay under the feet of White and Chaffee in order to swap the cable with another. An electrocardiogram reading at this time indicates he was engaged in some form of mild activity, possibly with the assistance of White, who could be seen on television monitors removing his glove. Brushing and tapping noises could be heard from within the spacecraft. The cable would later be found in a disconnected state, which could not have happened by accident.

At 6:30:55 p.m., something happened inside Spacecraft 012. Ground instruments monitoring the command module's systems and environment unexpectedly recorded a two-and-a-half-second interruption of power on an alternating current bus. At the same time, other monitors showed a sudden spike in the oxygen flow into the men's space suits. White's heart and respiration rates suddenly shot up. It seemed that a brief electrical arc suddenly flared between two bare segments of wire, believed to have been in a panel below the left-hand side of Grissom's couch and far removed from where he had been working on the cable.

Nine seconds later, flames appeared inside the spacecraft. Grissom yelled out what sounded like "Hey!" He scrambled up and knelt on his couch, banging his helmet hard on the upper instrument panel, leaving deep gouges in the top of the helmet. By now, clear oxygen-fed flames were sweeping up the inside wall of the cabin. Chaffee's voice suddenly broke through on the intercom, saying, "Fire – I smell fire." At 6:31:06, White's voice, this time far more terrifying, was heard to call, "Fire in the cockpit!" At this point he disconnected his oxygen inlet hose in order to do battle with the inner hatch release.

With everything in the cabin saturated with pure oxygen at high pressure, the fire rapidly consumed a host of combustible materials. Meanwhile the pressure had risen alarmingly and the crew was frantically going through their initial evacuation drills.

Chaffee turned up the lights and opened communication links. Ten seconds later he yelled in despair, "We've got a bad fire – let's get out – let's open her up!"

NO WAY OUT

Emergency escape procedures said that the hatches could be opened and the cabin evacuated in an orderly fashion in ninety seconds, but in simulations the crew had never achieved anything close to this time. In the evacuation exercises Grissom's role was to lower White's headrest so that White could reach above and behind his left shoulder to actuate the ratchet device that would simultaneously loosen the six dog-leg latches. The hatch itself was monstrously heavy, and opened inward. As shown on television monitors, White could be seen inserting the ratchet tool into a slot in the hatch. He suddenly snatched his hands back and then reached out once again as Grissom's hands also came into view, in a desperate attempt to help White with the hatch. Meanwhile the flames, which were mostly on Grissom's side of the cabin, rapidly grew in intensity, releasing poisonous gases that quickly suffocated the three astronauts.

Some recent research, combined with the independent findings of some NASA engineers, indicates that Grissom also tried to purge the pressure by thrusting his gloved hand through the flames in an attempt to activate the cabin dump valves on a shelf over the left-hand equipment bay. He pressed so hard and so violently that the valves were later found bent. However the valves did not fully engage. In any case, it is doubtful they would have had much effect on the rapidly mounting internal cabin pressure. Meanwhile, the temperature inside the spacecraft had grown high enough to melt stainless steel fittings. Molten balls of nylon were dripping onto everything. White's safety harness was on fire.

According to one source, White made part of a full turn of the ratchet before he was overcome by the deadly fumes, although it seems probable that the heat would have caused the metal in the hatch to expand and jam. It is widely acknowledged that White put in a mighty but futile effort to open the inner hatch. The last transmission from the spacecraft was a sharp, unidentified cry of pain.

In the meantime, the pressure inside the cabin had escalated to 36 psi, causing a sudden, violent rupture in the spacecraft's hull from a position adjacent to Chaffee's helmet across to below his feet. This explosion sealed the crew's fate. Accompanied by a howling roar, fierce flames and debris spewed out of the breach into the White Room, and fire briefly enveloped the outside of the command module. As the pure oxygen environment in the cabin rapidly depleted, the clear flames deepened in color and lost much of their intensity, replaced by a thick, dark, choking smoke.

From start to tragic finish the fire inside Spacecraft 012 only lasted about 14-17 seconds. Once the violent hull rupture had purged the cabin of oxygen, the fire was essentially extinguished.

Five minutes after the alarm had been raised the booster cover cap was opened, followed soon thereafter by the inner and outer hatches. At first, no one could see through the thick, swirling smoke, and there were no signs of activity from the crew.

Almost another five minutes would pass before the smoke had cleared sufficiently to reveal the inert bodies of the crew. Chaffee, badly injured in the explosion, was still strapped in his seat, while White had collapsed across his seat after several frantic efforts to open the hatch had failed. Even White, said to be among the fittest of the astronauts and as strong as an ox, never really had any chance of opening the hatch in time.

The grim aftermath of the fire showing the smoke-blackened hull of Spacecraft 012 after the protective booster cover cap had been removed. (Photo: NASA)

Grissom was found lying on his back on the floor of the spacecraft, where he had apparently crawled in an attempt to escape the fire. All three had their visors closed. The bodies of Grissom and White were so intertwined below the multilayered hatch that it was difficult to tell them apart. Doctors Fred Kelly and Alan Harter conducted a brief examination of the occupants and pronounced what everyone had known they would find – all three men were dead.

The only thing that could have saved the lives of the three astronauts would have been a fast-opening hatch.

A NATION IN MOURNING

Along with his crewmate Roger Chaffee, Gus Grissom was given a hero's burial at Arlington National Cemetery, Virginia on 31 January 1967 in a service which was broadcast nationwide on television. That same day, Ed White was buried with full military honors in West Point's Old Cemetery in Orange County, New York.

Their names are engraved on the Space Mirror Memorial at the Kennedy Space Center in Florida. At the request of the deceased astronauts' families, the mission that never flew was named Apollo 1, and when Neil Armstrong and Buzz Aldrin landed on the Moon in July 1969 they took with them an Apollo 1 mission patch which they left there.

Just days before he died, Gus Grissom finished drafting his book *Gemini: A Personal Account of Man's Venture into Space*. In it, he once again emphasized, "The conquest of space is worth the risk of life."

Immediately after the disaster, an Apollo 204 Review Board was established by NASA and this published its findings in April 1967. The fire was attributed to an electrical fault within the cockpit. Exacerbating factors were the extensive use of flammable materials in the construction of the cockpit and in the astronauts' space suits, plus the high-pressure pure oxygen atmosphere within the cabin. In response, NASA vastly reduced the amount of flammable materials and changed to a nitrogen-oxygen mixture in the cabin for ground testing through to launch. The agency also moved to a Block II command module design, the hatch of which could be opened far more easily and rapidly in the event of an emergency.

I asked former Mercury recovery helicopter pilot Jim Lewis where he was on the day of the Apollo pad fire, and for his reaction to the loss of Gus and the other two astronauts. "As I mentioned earlier, I was support team leader for Apollo 9, and we were conducting a closed hatch test on the Apollo 9 command module at the North American facility in Downey, California. The test was stopped, the hatch opened, and Jim McDivitt, the mission commander, was called away to the phone. He returned and told us what had happened. My reaction, and I think that of us all, was stunned, misty-eyed disbelief. Gus was one of the best and it just didn't seem possible at the time that any such thing could have happened. I was subsequently assigned to the hatch re-design team led by Frank Borman. As I recall, that took about six months and most of it was spent at the Downey facility. Those were good days also because all of us went about that task with a vengeance to insure that kind of thing would be precluded in the future."[13]

It took more than a year and a half before NASA was once again prepared to send astronauts into space, by which time the Apollo spacecraft had undergone extensive improvement. On 11 October 1968, Apollo 7, commanded by Gus Grissom's friend Wally Schirra, completed 163 orbits of the Earth during an eleven-day mission in the redesigned command module.

Deke Slayton always argued that the first person on the Moon should have been one of the original Mercury astronauts, and that man he had determined was Virgil Ivan ('Gus') Grissom. In his autobiography Slayton wrote, "One thing that would probably have been different if Gus had lived: the first guy to walk on the Moon would have been Gus Grissom, not Neil Armstrong."[14]

Norman Grissom is one who believes that his brother, had he lived, might have been selected as the first person to walk on the Moon. As he observed in 2000, "I think history has shown he was one of our outstanding people. He made a great contribution to the space race. Without him, I don't think the space program would be where it's at now."[15]

The crew of Apollo 7: Donn Eisele, Wally Schirra and Walt Cunningham. (Photo: NASA)

NASA astronaut Virgil ('Gus') Grissom. (Photo: NASA)

References

1. NASA MSC *Space News Roundup*, article "Project Mercury Announces Plans for Two-Man Rendezvous Spacecraft," issue Vol. 1, No. 4, (13 December 1961), pp. 1 & 7
2. *Missiles and Rockets* magazine, article "First Gemini – Grissom and Armstrong?" issue 17 February 1964, pg. 9
3. Borman, Frank and Robert J. Serling, *Countdown: An Autobiography*, Silver Arrow Books, New York, NY, 1988
4. Grissom, Virgil I., "Green on Gemini" speech before the U.S. Air Force Academy, Colorado Springs, Colorado, 25 January 1963

5. Grissom, Virgil, *Gemini: A Personal Account of Man's Venture Into Space*, The Macmillan Company/World Book Encyclopedia, Toronto, Canada, 1968
6. *Ibid*
7. *Ibid*
8. Grissom, Virgil I., "Green on Gemini" speech before the U.S. Air Force Academy, Colorado Springs, Colorado, 25 January 1963
9. Slayton, Donald K. with Michael Cassutt, *Deke! From Mercury to the Shuttle*, Forge Books, New York, NY, 1994
10. *Missiles and Rockets* magazine, "The Countdown" page article "Grissom Moves Toward Apollo CP Role", issue 14 February 1966
11. Grissom, Virgil, syndicated article, *Three Times a Command Pilot*, November 1966
12. Barbour, John and the writers of the Associated Press, *Footprints on the Moon*, The Associated Press, 1969, pg. 125
13. E-mail correspondence James D. Lewis with Colin Burgess, November 2002 to August 2003
14. Slayton, Donald K. with Michael Cassutt, *Deke! From Mercury to the Shuttle*, Forge Books, New York, NY, 1994
15. Grissom, Norman, interviewed by John Norberg of Associated Press for article, "Indianapolis Museum to Honor Indiana Astronaut," 14 October 2000.

8

Epilogue: From the depths of the ocean

Curt Newport cannot recall when the idea first occurred to him to consider the possibility of raising *Liberty Bell 7* from the ocean floor. "It might have been when I read *The Right Stuff*, or it could be just something I thought of," the salvage operator ventured during an interview back in 1986, a full quarter of a century after the loss of Gus Grissom's spacecraft. All he knew back then was that it had sunk in very deep water and that any recovery effort would be an incredibly difficult task.[1]

A CHILDHOOD FASCINATION WITH SPACE

He was born in Oakland, California, where his father flew as an Army aviator out of Chrissy Field. Growing up with a childhood passion for space flight and undersea exploration, Curt Newport was only 10 years old and living in St. Louis, Missouri, where his father was stationed temporarily, when *Liberty Bell 7* was lost on 21 July 1961, settling into the mud of the Florida Trench off the Bahamas, some three miles below the surface of the Atlantic. "I think Grissom's capsule was probably built less than ten miles from our home," he reflected in 2013.[2]

As he related in his book, *Lost Spacecraft: The Search for Liberty Bell 7*, the Mercury astronauts were huge heroes back then. "While Shepard and Glenn were certainly the most famous to me, I remember being taken by Grissom for no special reason. Maybe it was the way he looked or that he didn't appear to seek out the limelight. However, he was a central figure to me."[3]

In 1974, aged 24, Newport entered into the subsea business building ship fenders in Washington, D.C. He later graduated into building deep diving systems such as diving bells and deck decompression chambers. After leaving a Los Angeles-based commercial diving school in 1977 he began working with submersible robots known as Remotely Operated Vehicles (ROV). Although his expertise grew over time, he found much of the freelance work in which he was engaged, such as inspecting rusty

C. Burgess, *Liberty Bell 7: The Suborbital Mercury Flight of Virgil I. Grissom*, Springer Praxis Books, DOI 10.1007/978-3-319-04391-3_8, © Springer International Publishing Switzerland 2014

pipelines and routine maintenance on AT&T telephone cables, to be rather less than satisfying. He began to look at people involved in ocean exploration such as Jacques Cousteau for some way to creatively inspire and challenge him. "I was interested in doing something that *I* felt was worthwhile with the underwater vehicles that I had worked with for so many years. I wanted to have some fun with ROVs."

Newport says he had very little money back then, but a lot of ideas. "I started thinking about things that had been lost in the ocean. *Targets.* Sunken objects that would be interesting to find and explore and I came up with two possibilities – the *Titanic* and Gus Grissom's *Liberty Bell 7* Mercury spacecraft."[4] In an article which he wrote before *Titanic* was located, he actually predicted the likely location of the ocean liner to within a couple of miles.

In 1985 he was contracted to remotely pilot the SCARAB 2 ROV, equipped with television cameras, sonar, and mechanical arms to help salvage the wreckage of an Air India 747 airliner off the coast of Ireland. A total of 329 people, including 268 Canadians, died *en route* from Montreal to New Delhi when the aircraft was ripped apart 31,000 feet above the Irish Sea by a bomb which was planted on board by the Sikh militant group Babbar Khalsa. It remains the deadliest aviation disaster ever to occur over a body of water.

"AI 182 was actually found by a Navy search team using a towed pinger locator and side-scan sonar before I arrived in Ireland in July of 1985," noted Newport. "By the time I got there onboard the CCGS *John Cabot*, Cable and Wireless had already recovered the FDR [flight data recorder] and CVR [cockpit voice recorder] using SCARAB I. What I did was survey the crash site, a three-by-five nautical mile area, and recover wreckage using SCARAB II in conjunction with a German ship which had all the heavy lift gear. The data recorders proved nothing. But evidence of an explosion was on the wreckage we raised. We broke lots of records on those dives, one lasting 143 hours." Altogether, the exhausting salvage operation continued for six months, ending in November 1985.[5]

FURTHER INVESTIGATIONS

The Air India experience caused Newport to ponder further the difficult question of locating and salvaging the lost Mercury spacecraft. As he researched where it might be on the ocean floor, he realized that no existing ROV was capable of reaching an object three miles down. After returning from Ireland he asked the Smithsonian's National Air and Space Museum (NASM) about *Liberty Bell 7* "but got nowhere."

His next major salvage assignment involved the tragic loss of NASA's space shuttle *Challenger* and her crew of seven on 28 January 1986. Newport spent two months working out of Port Canaveral, Florida on the contract salvage ship *Stena Workhorse*. The major discovery of the operation occurred while Newport was on a midnight shift piloting the Gemini ROV and brought a booster section to the surface. It happened to be the most crucial find of the search – the segment of the right-hand solid rocket booster where the burn-through of an O-ring had set in motion the fatal explosion and NASA's greatest tragedy to that time.

A 4,000-pound segment of *Challenger*'s right-hand solid rocket booster is offloaded at Port Canaveral from the *Stena Workhorse* following its recovery on 13 April 1986. (Photo: NASA-JSC)

"The operation was a real grind," he says, "mostly due to the numerous technical problems we had with the Gemini ROV. During six weeks we repaired its electrical umbilical a staggering 32 times and even replaced the whole thing four times: not a good record. But while I was in Florida and on one of my rare days off, I visited the archives at the Kennedy Space Center and collected a little more data on *Liberty Bell 7*."[6]

He also began to establish solid contacts in his ongoing research into the loss of Grissom's spacecraft, including the Gemini and Apollo astronaut Tom Stafford, who would not only prove to be a staunch advocate of Newport's plans but also provided important leads and privileged access to documents and information to assist him in his quest to pinpoint the location of the sunken craft. "Gene Cernan, John Yardley (McDonnell Aircraft Corporation) and Robert F. Thompson (JSC) were also a big help."[7]

Another interested and influential ally was Max Ary, then President of the Kansas Cosmosphere and Space Center in Hutchinson, Kansas, who had also considered the possibility of finding and recovering *Liberty Bell 7*. As he told Lawrence McGlynn for collectSPACE, "Actually my interest, relative to the Cosmosphere, in recovering the *Liberty Bell 7* goes way back. In 1978, before the Cosmosphere opened but when we were still trying to put together a space artifact collection, one of my many basic goals was to place on exhibit examples of *all three* of the early manned spacecraft. We knew when we were going to get the Gemini and Apollo, but we knew, because of the rarity of the Mercury [capsules] that *it* was going to be our biggest challenge. When I realized that all of the available Mercury [craft] were on long-term exhibit, it occurred to me there was still one that might be made available, and that was LB7.

Being from Kansas, and with no knowledge of the ocean, I didn't specifically see why there would be any problem in recovering something from 16,000 feet down. As they often say 'ignorance is bliss.'"[8]

A MAJOR BREAKTHROUGH

In 1987, following further ROV recovery operations, Newport took on a position with Oceaneering Space Systems, which involved working on the Space Station *Freedom* program. By this time he had established a fairly good grasp on where *Liberty Bell 7* might be located, after poring over countless documents and charts. Then he had a major breakthrough.

"While working at Oceaneering Space Systems, I learned that they were planning to do some deep water sea trials using the Gemini ROV we'd used on the *Challenger* salvage. It had been updated and now had a 15,000 foot depth capability, so I made the suggestion: Why not add a side-scan sonar to the trial and use the opportunity to look for *Liberty Bell 7*? After considerable back and forth with several Oceaneering vice presidents, they decided to give it a try using Steadfast Oceaneering's Deep Ocean Search System (DOSS)."[9]

The trial eventually went ahead, and the search was conducted in the area where Newport reasoned that the capsule might reside. There was excitement when two objects – one large and one small – were located, but in a curious twist of fate they later turned out to be pieces of wreckage from a downed aircraft. After several years spent scouring NASA charts and photographs and interviewing those present when *Liberty Bell 7* went down, Newport remained undiscouraged. A thorough check of weather and sea conditions on the splashdown day in 1961, as well as currents in that section of the Atlantic, led him to the conclusion that *Liberty Bell 7* did not drift far before sinking. He also believed that despite the massive pressure at that depth, the capsule would have remained basically intact. The only real uncertainty he harbored was whether it had moved horizontally during its nearly hour-long fall to the ocean floor. Nevertheless he was convinced he could locate the spacecraft, and mounted two further ROV expeditions in 1992 and 1993. But these were ancillary ventures attached to other seaborne operations, and were conducted in haste.

As he commented to the author, "Actually I was discouraged much of the time and gave up on the project during certain periods. You should see all the rejection letters I have. I was very concerned about the SOFAR [Sound Fixing and Radar] bomb carried in the spacecraft, even though there was no evidence it detonated – but it should have."[10] The SOFAR device was designed to go off at a depth of 3,000 feet if the spacecraft sank, allowing recovery vessels to pinpoint its location.

Newport continued to work with ROVs on various salvage projects, including the recovery of wreckage from yet another downed airliner. On 17 July 1996 the 747 on flight TWA 800 had mysteriously exploded and crashed into the Atlantic near East Moriches, New York. This time the probable cause was an explosion in a fuel tank sparked by a short circuit. In the first two weeks on the TWA operation, Newport's team recovered the bodies of over 50 passengers using the Navy's MR-1 ROV.

Over the years, Newport had participated in the development and use of ROVs and knew they were now far more reliable and easier to mobilize. "Overall, by 1998, things were looking up for me," he recalled. Then he heard that Oceaneering might be conducting some dives on the RMS *Titanic* for the Discovery Channel, and he became part of the team, this time in charge of remote-piloting an ROV known as Magellan. "I actually got MSNBC and Discovery the 'promo' which they used to advertise the [*Titanic*] program by flying the Magellan straight up the edge of the bow, very close and very fast – so close that I knocked off rusticles [formations of rust similar in appearance to stalactites] from the towing shackle with the priceless WHOI [Woods Hole Oceanographic Institution] high-resolution camera. A little too close I guess, but that's what they wanted."[11]

Prior to this expedition, he had written to the Discovery Channel in regard to his own near-quixotic quest to locate *Liberty Bell 7*. To his surprise, he was aboard the ship *Ocean Discovery* one day when he got a life-changing call from the Discovery Channel's Tom Caliandro. After discussing the project it was agreed that a meeting would take place once he returned from his work on the *Titanic*.

"Discovery had actually already turned me down in the early 1990s regarding *Liberty Bell 7*. The only reason I wrote them again was at the urging of a friend of a man doing renovation work on our house; he wanted to break into documentary film making. I never expected anything to come of it. Then the next thing I knew, I was getting phone calls in Boston while mobilizing the Magellan 725.

"What happened, is that after the *Titanic* operation I came back home from Newfoundland because I was scheduled to do classified work for the Navy in England within a week or so. During my three days home before flying out, I met with Discovery in Bethesda, Maryland, and wrote a business plan which was delivered to Discovery while I was on my way to England. I think they gave final approval to the project early in 1999."[12]

FIRST EXPEDITION

Curt Newport's team set sail on a two-week voyage on Monday, 19 April 1999. This time the Discovery Channel was paying for the entire expedition as well as filming the venture for a documentary to be broadcast in the fall. Everyone was hoping for a successful conclusion. The ship, the MV *Needham Tide*, was equipped with the very latest in side-scan sonar unit, although in reality the ship was barely suitable for a sonar search because of its propulsion system; it didn't even have variable pitch propellers. The only way they could operate it sufficiently slowly to tow at 1.5 knots was to put one screw ahead and one astern, which they did for a whole week.

"We have a pretty good idea where to look for it," Newport said prior to sailing. "To say I'm cautiously optimistic is probably the right term … this is a full-fledged, dedicated mission to go out and locate and recover this thing. We have a lot more time, we have a better sonar, we can examine a much larger area of the ocean at one time. Consequently, our chances are better. Outside of *Challenger*, this is the only one we haven't gotten back. It's the right thing to do. It's patriotic. It was one of ours."[13]

The Ocean Explorer 6000 side-scan sonar which was dragged over the bottom of the Atlantic in the target area and identified 88 sonar contacts. (Photo courtesy of Curt Newport)

Curt Newport (left) reviews the 88 target sheets generated by the sonar search for likely contacts. (Photo: Discovery Channel, courtesy of Curt Newport)

Misfortunate quickly set in when the side-scan sonar broke down, but once fixed it identified 88 possible targets scattered across the 24 square mile area that Newport had pinpointed based on 14 years of studying NASA charts and interviewing people who were there when *Liberty Bell 7* went down, such as helicopter pilot Jim Lewis.

Launching the Magellan 725 ROV. (Photo courtesy of Curt Newport)

On Saturday evening, 1 May, on Day 14 of the expedition, something bizarre happened. At 7:20 p.m., on the very first attempt and to everyone's amazement, a vaguely familiar shape – a veritable ghostly apparition – emerged from the murky terrain as they stared at the video screen. Newport would not allow himself to get carried away; his first reaction was that it might be more large pieces of aircraft wreckage. After all, Target #71 was the very first object the Magellan Rover had closed in on. But as it drew nearer there was no mistaking the bell-shaped object with its rough exterior; it was the Mercury capsule *Liberty Bell 7*. Everyone, and especially Newport, was stunned at hitting the target first time.

"Everyone said that we simply got lucky. But it was not luck that we were looking in that area. Target 71 was one of a cluster of five hard contacts centered near my wind-corrected GBI FPS-16 radar location. That target was the first we looked at because it made sense from the operational standpoint, that is, from the positioning and navigational standpoints. In reality, the first target we found was a trail of LB7's decomposed heat shield, which was scattered down a small rise from the capsule. We had no idea what we were looking at. I thought that it was aircraft debris. The next thing I knew, we were seeing this tall thing in the darkness up the hill."[14]

From what they could see, the spacecraft appeared to be intact and the words *Liberty Bell 7* could be clearly discerned on the side of the craft. And so could the false, painted crack. As Newport later pointed out, the spacecraft could easily have been obscured from the sonar and they might not have seen it in the gloom. Good fortune was certainly riding with them that day since it was a very small object in a massive area of ocean floor. "What we did find was something in water half a mile deeper than the *Titanic* and it's smaller than one of the *Titanic*'s boilers."[15]

Sadly, the team's elation would not last long. At four minutes to midnight the cable to the remotely operated Magellan 725 ROV snapped in the rough seas and the rover

After five hours probing the depths, the Magellan ROV located some mysterious glistening chunks of white material Newport described as "space age bread crumbs" which led the team to *Liberty Bell 7*. (Discovery Channel image courtesy of Curt Newport)

The first ghostly images of *Liberty Bell 7* captured in Magellan's powerful lights. (Discovery Channel image courtesy of Curt Newport)

was lost, sinking to the sea floor 15,800 feet below the *Needham Tide*. It meant the team would have to abandon the operation and return to Port Canaveral to replace the rover, which would set them back a few weeks. But now they knew precisely where the capsule was. Despite the loss of an expensive ROV, Newport maintained a sense of optimism in a call he made from the ship. "It looks to be in beautiful condition," he reported of the spacecraft, "and certainly capable of being recovered."

The lost ROV would not remain on the ocean floor for long. "The Magellan 725 was later recovered about a week after we returned to the Cape with the capsule. I personally believe that Oceaneering would've left it there except for the expensive camera we had leased from WHOI. Woods Hole wanted that camera back, which was the same one we'd used to image the *Titanic* the previous year. They left the depressor and several thousand feet of armored optical fiber cable on the bottom simply by cutting the soft tether and lifting the ROV."[16]

It would take Oceaneering International five weeks to construct another ROV, this time named Ocean Discovery. The expedition team sailed again on 4 July on a different ship, the *Ocean Project*. They stopped briefly at Fort Lauderdale to pick up a few people, including Jim Lewis, the helicopter pilot who had unsuccessfully attempted to rescue the sinking capsule in 1961, and Guenter Wendt, McDonnell's former launch pad leader. Two bomb experts from UXB International were also on board to deactivate the explosive SOFAR navigation device which had apparently failed to detonate when the spacecraft sank.

FINDING A LOST SPACECRAFT

On the evening of Wednesday, 20 July 1999, *Liberty Bell 7* was finally recovered from the depths of the Atlantic, just one day short of the 38th anniversary of Gus Grissom's suborbital MR-4 mission, and 30 years to the day that Neil Armstrong and Buzz Aldrin had walked on the Moon.

It took eight hours to bring the spacecraft to the surface. "I actually designed the recovery tools used to create a new lift point on the capsule as the old Dacron loop was judged not suitable," Newport explained. "This was done while I was working at Oceaneering Space Systems and the dimensions were checked using the MA-1 wreckage that Max [Ary] had at the Cosmosphere (in 1993). After finding the capsule and before the second trip, I took one of the tools to NASM's Garber facility and established where we could and could not attach them on the escape tower mounting ring. You just couldn't place them anywhere, as fasteners and sensors interfered in certain areas. I used the images I took while the tools were being installed during the recovery to make sure they were being put where they should be. Oceaneering had earlier misplaced three of the four tools but I had one of them in my bedroom closet for years, always hoping that I'd get a chance to use it."[17]

Lifting *Liberty Bell 7* was a delicate operation. The water-filled craft weighed 3,000 pounds as it was slowly raised to the surface, so even using a super-strong Kevlar line everyone proceeded with the utmost of caution. To the former *Hunt Club 1* helicopter pilot Jim Lewis's amazement, when *Liberty Bell 7* finally broke to the surface he saw that the recovery line which his co-pilot John Reinhard had attached in 1961 was still dangling from the top of the capsule. As he commented later, "To see it come out of the water again, like it did that long ago, was a feeling that I don't have an adjective to apply to." [18]

Once the spacecraft had been successfully raised and anchored onto the deck of the ship, the two bomb experts carefully rummaged around the capsule's interior and removed the explosive SOFAR device, which they threw overboard.

Considering the time that the capsule had spent in the murky depths, it was in surprisingly good condition as Newport noted, although the beryllium heat shield had totally disintegrated. "I can still see the actual straps that Grissom wore during his flight," he reported during a phone call from the ship. "The personal parachute inside the spacecraft is perfectly intact." But things made of aluminum, such as the control panels, had deteriorated badly.[19]

Newport would later write: "While *Liberty Bell 7*'s condition was indeed remarkable, the interior was a mess. The forward hatch had come loose from its mount and was lodged in the explosive hatch opening. I gingerly extracted it from the capsule and finally had my first look at the interior. Water continued to drip all through the inside as I stuck my head in, seeing that part of the once intact control panel had disintegrated, leaving numerous flight instruments dangling like apples on a tree. But what struck me most was the smell ... It was like the odor of carbon or decayed wood, probably from the chemical action of the electrolyte in the craft's batteries. The optical periscope had broken in half and was now laying amongst other rubble such as the decomposed remains of the control panel and what was left of the astronaut camera. I could see part of one of the film spools lying exposed, which eliminated any hope of saving the film."[20]

When asked if they had also managed to locate the spacecraft's hatch, Newport responded that due to a two-day delay caused by problems with the navigational data they had no time to search for the hatch, and the important thing now was to get the spacecraft to port as soon as possible. Once there the Kansas Cosmosphere and Space Center people would transport *Liberty Bell 7* to Hutchinson, Kansas, where their experts would disassemble and clean the capsule.

It was thought possible that the lingering mystery of why the hatch blew might have been recorded in the cockpit film camera that was running as *Liberty Bell 7* splashed down, but as Newport had discovered the camera had broken open and the film was completely ruined. "I don't think there's going to be any way to answer that question, ever," Newport said at the time. He added that he had no intentions of going back to look for the hatch, ever. "Finding the spacecraft was good enough."[21]

Time would change Newport's opinion on recovering the hatch. "Well, they say never say never, and I don't remember exactly what I said regarding the hatch.

One item later recovered from within Liberty Bell 7 was the explosive hatch igniter knob. (Photo: Kansas Cosmosphere and Space Center)

But I now think it can be found and *is* worth recovering. It was always in the plan to look for the hatch and I already had three targets worth investigating. At this point, I'm looking for money and equipment to go back to the site."[22]

A SPACECRAFT FINALLY LANDS

Following the spacecraft recovery, Newport and the Kansas Cosmosphere's Max Ary began their documentation and post-recovery work. After taking photographs, Newport used a bilge pump to remove pooled water from the bottom of the capsule while Ary reached through the hatch to see what he could find. To their amazement he felt something unusual and upon withdrawing his hand from the muck revealed several shiny Mercury dimes.

To prevent further degradation and corrosion following the capsule's exposure to the air, *Liberty Bell 7* was placed into a specially designed container filled with sea water for shipment to the mainland – to a place not far from where it had ascended into the sky atop a Redstone rocket 38 years before. "Putting that capsule into the container at night, and at sea, was a harrowing experience. We had it in a cargo net and against my explicit instructions the crane operator slewed the capsule around over the sea off the port side with the ship rolling like crazy. I literally put my body between the capsule and the ship's side to keep them from colliding and *Liberty Bell 7* was dunked several times in the water to stabilize things. But we got it done and no one lost any fingers while installing the heavy container top."[23]

As the *Ocean Project* neared Port Canaveral, Curt Newport called the recovery crew together and thanked them for their involvement in the incredibly successful mission – the deepest commercial salvage operation in history.

"I popped the cork on a magnum bottle of Moët champagne and we celebrated as much as our one bottle allowed."[24]

Once on board the *Ocean Project*, *Liberty Bell 7* was placed in a special water tank to keep the spacecraft moist in the hot sun. (NASA-KSC, Photo ID KSC-99PP-1033)

FINAL THOUGHTS

Curt Newport is not one to let the grass grow under his feet. Since the recovery of *Liberty Bell 7* he has been involved in a number of enterprises and recoveries. These include the recovery of Blackhawk helicopter 221 off Fiji; the Australian Navy's Remora Submarine Rescue System near Fremantle, Western Australia; the search for Air France 447 out of Natal, Brazil; and locating an E-2C Hawkeye (Bluetail 601) in the Arabian Sea and an AV-8 Harrier in the Gulf of Aden. More recently there was a

Curt Newport talks to reporters on the dock at Port Canaveral. (NASA-KSC, Photo ID KSC-99PP-1030)

Former McDonnell pad leader Guenter Wendt (left) and Lowell Grissom view the salvaged spacecraft. (Photo: Kansas Center and Space Center)

Republic of China Air Force F-16 fighter that went down in the Taiwan Straits and a highly classified mission which, unfortunately, no one will ever know about.

"People never know much about my 'real job,'" he muses, "and based on *Liberty Bell 7*, the *Indianapolis*, or the *Belgrano*, probably believe I've had few successes. But in reality, most of the search and recovery operations I've supported have been successfully completed. It helps to have the government's money to do them right; most of these historic project are always done cutting corners to save money and that is while many of them fail."[25]

In concluding the recovery story, Curt Newport asked to make one final comment. "In the end, after the dust settled, it was good to be proven right and everyone else wrong. It simply came down to the fact that no one else knew what I knew because they had not read all the documents, done the underwater work in the ocean, and simply thought of a way to do it. There was never any question in my mind that it could be done. All we needed was the will to do it, the right equipment, people, and of course, the money.

"The only bad taste in my mouth about this thing is that NASA, as an agency, never said an official word to me after the recovery, unlike Jeff Bezos' F-1 engine salvage [in 2013]. I did receive a couple of nice letters from George Abbey (JSC) and Art Stephenson (Marshall Space Flight Center), but that was only because they knew me and recognized how much work I did. But NASA? Never a word. I think they simply regarded the capsule's loss as an embarrassment, even after 38 years."[26]

A RESTORATION TASK LIKE NO OTHER

In researching this book, I happened across an extremely well written article in the Spring 2000 (Vol.7, No.4) issue of the quarterly history of space flight magazine, *Quest*, penned by Connecticut-based space writer Keith Scala. Under the title, *The Future of Liberty Bell 7*, his article related the manner in which the spacecraft had undergone meticulous restoration at the Kansas Cosmosphere and Space Center and the plans for its subsequent exhibition. I contacted Keith to ask if he might consider updating his article as a feature in this book, to which he happily agreed. Here then, is the post-recovery story of *Liberty Bell 7*.

THE RESTORATION OF *LIBERTY BELL 7*

Keith J. Scala

After *Liberty Bell 7* was removed from the expedition ship *Ocean Project* at Port Canaveral, the capsule was transported to the Kansas Cosmosphere and Space Center in Hutchinson, Kansas on 1 September 1999. The capsule was shipped overland in a container filled with seawater to prevent further corrosion

The Kansas Cosmosphere is the only private museum authorized by the Smithsonian Institution to restore U.S. manned spacecraft. One previous spacecraft restored there in 1997 was the Apollo 13 command module *Odyssey*. Several other spacecraft from the United States and Russia have also been restored at the Midwestern facility.

Liberty Bell 7 arrives at the Kansas Cosmosphere. (Photo courtesy of the Kansas Cosmosphere and Space Center)

Max Ary, the Kansas Cosmosphere's president, explains the restoration process the spacecraft would undergo. (Photo: Kansas Cosmosphere and Space Center)

In the case of *Liberty Bell 7*, the Kansas Cosmosphere was given ownership of the spacecraft. In fact, *Liberty Bell 7* is the only manned American spacecraft not owned by NASA or the Smithsonian. Since NASA and the Smithsonian had never planned to recover *Liberty Bell 7*, it was agreed that the Kansas Cosmosphere would retain ownership of the ill-fated capsule. Before the successful recovery operation NASA gave ownership to the Discovery Channel. After recovery the ownership was then transferred to the Kansas Cosmosphere. This was done to help reimburse the cost and time of the recovery effort.

The intention of the Kansas Cosmosphere was to clean every portion of the spacecraft and remove parts that had been so badly damaged by corrosion that it would not be feasible to restore them. Max Ary, past president and CEO of the Kansas Cosmosphere, said at the time, "We will disassemble the entire spacecraft, individually cleaning every single piece, from the exterior shingles to the smallest of screws. Our goal is not to make the spacecraft look brand new," Ary added. "We simply want to clean and preserve it so that it will be available for generations to come. We will reassemble the spacecraft with only the original parts. If we need a particular part (the original part was too badly damaged), for example, a panel into which switches need to be mounted, we will create one of Plexiglas or we will create a metal skeleton that will serve the same purpose. Either way, it will be obvious to the viewer which pieces of *Liberty Bell 7* did not survive the 38 years in a harsh deep-sea environment." The Plexiglas or framework has the added benefit of allowing the viewer to see the interior of the spacecraft not normally seen. The cleaning also stabilized the capsule so no additional deterioration would occur after the restoration.[27]

The overall condition of *Liberty Bell 7* was surprisingly good after its 38-year stay at the bottom of the Atlantic. First, 75 gallons of mostly sand and other things that live in the ocean were removed from the capsule. Many metal parts, mostly those made of aluminum, were excessively corroded while items like paper and plastic had been preserved. The superstructure of the capsule was intact. The capsule control panels had corroded away but many switches and gauges were in excellent condition. Some glass face plates on the gauges had been shattered by the water pressure.

The capsule structure has two layers; firstly, an outer skin of *René 41* alloy shingles used to protect the capsule from reentry heat. The shingles are bolted to the inner titanium structure with aircraft-style stringers in between. This method provided for a gap between the outer and inner structures. The gap allowed ceramic fiber to be used for extra thermal insulation. The inner structure serves as a pressure vessel to retain air pressure. The outer skin shingles were removed as well as all the interior equipment. The words UNITED STATES and even the "Liberty Bell" crack painted on the outer skin remained.

A harsh but necessary method had to be used after the outer alloy shingles were removed. Since technicians could not remove the lower bulkhead of the inner structure, as it was welded to the bottom of the conical inner structure, it was decided to cut the capsule into two pieces. The cut was made between the entry hatch and lower bulkhead around the diameter of the capsule.

This allowed access to each of the 24 framed aircraft-style stringers that run vertically inside. The lower half and top half of the capsule could then be sandblasted free of corrosion. When the capsule was reassembled and the outer alloy shingles were attached to the inner structure the cut would not be evident.[28]

Many interesting items were found inside the capsule. The film inside a camera that kept a record of Grissom's movements was not salvageable, but it was hoped that a magnetic tape with Grissom's voice during the flight could be saved. A roll of Mercury dimes intended to be given to Grissom's friends was discovered, several of which were revealed at the capsule's Port Canaveral homecoming. Several silver dollar certificates bearing Grissom's signature were also found. The dollar bills had been rolled up and deliberately hidden inside plastic shrink-wrap tubing to look like part of the interior wiring. More than likely technicians would have removed these after the flight (as described earlier in this book) as mementos of the flight.

A checklist and grease pencil Grissom used during the flight was still in useable shape. A metal cap that covered the explosive hatch detonator was found, but did not shed any light as to why the hatch blew, causing the capsule to flood. Curiously enough, a bar of Dial soap was still in its paper wrapper and in good shape, even though several metal parts of the capsule had corroded away. One last item of interest is a portion of Teflon cable attached to the top of the capsule. This cable had been used by the Marine helicopter in a vain attempt to pull the capsule from the water, but had to be cut when the capsule started to pull the helicopter down.[29]

The restoration process took approximately seven months and cost $250,000. It involved 7,280 manned hours and 30,000 parts, while 10 miles of wiring had to be removed from the spacecraft and replaced. During the restoration process a webcam broadcast pictures of the spacecraft 24 hours a day, allowing Internet viewers to

Liberty Bell 7 after removal of the outer shingles. Note the cut made to the capsule under the hatch. (Photo courtesy of the Kansas Cosmosphere and Space Center)

follow the restoration process. Visitors to the Kansas Cosmosphere were able to view the restoration of *Liberty Bell 7* from the other side of a Plexiglas wall.

After the restoration was completed, no definitive information was found on why the hatch mysteriously blew after splashdown. If the hatch itself had been found (it was never recovered) it might have given some answers. Had *Liberty Bell 7* not been lost, it also could have been helpful to the official investigation into the accident in 1961.

The Mercury head dimes found inside *Liberty Bell 7*. (Photo: NASA/KSC, ID KSC-99PP-1035)

A number of recovered Roosevelt dimes carried by Gus Grissom eventually found their way onto the collectors' market. (Photo: The Skyman1958 collection)

Liberty Bell 7 after restoration. (Photo courtesy of the Kansas Cosmosphere and Space Center)

When the restoration process was finished, the Discovery Channel took *Liberty Bell 7* on a tour of North America. An interactive 6,000 square-foot traveling exhibit was used with *Liberty Bell 7* as the centerpiece. The tour started in 2000 and lasted six years. After the tour was completed, *Liberty Bell 7*'s final home was the Kansas Cosmosphere. Should any future space historians wish to reopen the investigation into how the hatch blew back in 1961, *Liberty Bell 7* will be waiting in Kansas to help answer the mystery.

References

1. Reichhardt, Tony, article, "ISO: Liberty Bell", from *Space World* magazine, issue January 1987, pp. 25-28
2. E-mail correspondence Curt Newport with Colin Burgess, 28 October – 3 November 2013
3. Newport, Curt, *Lost Spacecraft: The Search for Liberty Bell 7*, Apogee Books, Ontario, Canada, 2002, p. 110
4. *Ibid*
5. E-mail correspondence Curt Newport with Colin Burgess, 28 October – 3 November 2013
6. Newport, Curt, *Lost Spacecraft: The Search for Liberty Bell 7*, Apogee Books, Ontario, Canada, 2002, p. 110
7. E-mail correspondence Curt Newport with Colin Burgess, 28 October – 3 November 2013
8. Ary, Max, interview with Lawrence McGlynn, 8 May 2003
9. Newport, Curt, *Lost Spacecraft: The Search for Liberty Bell 7*, Apogee Books, Ontario, Canada, 2002, p. 120
10. E-mail correspondence Curt Newport with Colin Burgess, 28 October – 3 November 2013
11. *Ibid*
12. *Ibid*
13. Dunn, Marcia, article "Salvage Expert, Discovery Channel to Search for Mercury Capsule," The Associated Press, 15 April 1999
14. E-mail correspondence Curt Newport with Colin Burgess, 28 October – 3 November 2013
15. Dunn, Marcia, article, "Gus Grissom's Sunken Mercury Capsule Found,", The Associated Press, 3 May 1999
16. E-mail correspondence Curt Newport with Colin Burgess, 28 October – 3 November 2013
17. *Ibid*
18. Dunn, Marcia, article, "Plucked From Deep, Mercury Capsule Retrieved," The Associated Press, 21 July 1999
19. *Ibid*
20. Newport, Curt, *Lost Spacecraft: The Search for Liberty Bell 7*, Apogee Books, Ontario, Canada, 2002, pp. 173-174
21. Dunn, Marcia, article, "Plucked From Deep, Mercury Capsule Retrieved," The Associated Press, 21 July 1999
22. E-mail correspondence Curt Newport with Colin Burgess, 28 October – 3 November 2013
23. *Ibid*
24. Newport, Curt, *Lost Spacecraft: The Search for Liberty Bell 7*, Apogee Books, Ontario, Canada, 2002, p. 177
25. E-mail correspondence Curt Newport with Colin Burgess, 28 October – 3 November 2013

26. *Ibid*
27. E-mail correspondence Jim Remar (President/COO, KC&SM) with Keith Scala, 23 October 2013
28. Technical restoration report of *Liberty Bell 7* Kansas Cosmosphere
29. You Tube video "Behind the Scenes with *Liberty Bell 7* Treasures" presented by Chris Orwoll (CEO, KC&SM), 18 April 2009

Appendix 1: Pilot Virgil I. Grissom's post-flight Mercury-Redstone (MR-4) report

(References to the accompanying slide presentation deleted)

INTRODUCTION

The second Mercury manned flight was made on July 21, 1961. The flight plan provided a ballistic trajectory having a maximum altitude of 103 nautical miles, a range of 263 nautical miles, and a five-minute period of weightlessness.

The following is a chronological report on the pilot's activities prior to, during, and after the flight.

PRE-FLIGHT

The pre-flight period is composed of two distinct areas. The first is the training that has been in progress for the past 2.5 years and which is still in progress. The second area, and the one that assumes the most importance as launch day approaches, is the participation in the day-to-day engineering and testing that applies directly to the spacecraft that is to be flown.

Over the past two years, a great deal of information has been published about the astronaut training program and the program has been previously described in Reference 1. In the present paper, I intend to comment on only three trainers which I feel have been of the greatest value in preparing me for this flight.

The first trainer that has proven most valuable is the Mercury procedures trainer which is a fixed-base computer-operated flight simulator. There are two of these trainers, one at the NASA-Langley Air Force Base, Virginia, and one at the Mercury Control Center, Cape Canaveral, Florida. These procedures trainers have been used continuously throughout the program to learn the system operations, to learn emergency operating techniques during system malfunctions, to learn control techniques, and to develop operational procedures between pilot and ground personnel.

During the period preceding the launch, the trainers were used to finalize the flight plan and to gain a high degree of proficiency in flying the mission profile. First, the systems to be checked specifically by the pilot were determined. These were to be the manual proportional control system; the rate command control system; attitude control with instruments as a reference; attitude control with the Earth-sky horizon as a reference; the UHF, HF, and emergency voice communications systems; and the manual retro-fire override. The procedures trainer was then used to establish an orderly sequence of accomplishing these tasks. The pilot functions were tried and modified a great number of times before a satisfactory sequence was determined. After the flight plan was established, it was practiced until each phase and time was memorized. During this phase of training, there was a tendency to add more tasks to the mission flight plan as proficiency was gained. Even though the MR-4 flight plan contained less pilot functions than the MR-3 flight plan, I found that the view out the window, which cannot be simulated, distracted me from the less important tasks and often caused me to fall behind the planned program. The only time this distraction concerned me was prior to retro-fire; at other times, I felt that looking out the window was of greater importance than some of the planned menial tasks. In spite of this pleasant distraction, all tasks were accomplished with the exception of visual control of retro-fire.

The second trainer that was of great value and one that I wish had been more readily available prior to launch was the air-lubricated free-attitude (ALFA) trainer at the NASA-Langley Air Force Base, Virginia. This trainer provided the only training in visual control of the spacecraft. I had intended to use the Earth-sky horizon as my primary means of attitude control and had spent a number of hours on the ALFA trainer practicing retro-fire using the horizon as a reference. Because of the rush of events at Cape Canaveral during the two weeks prior to launch, I was unable to use this trainer. I felt this probably had some bearing on my instinctive switch to instruments for retro-fire during the flight, instead of using the horizon as a reference.

The third training device that was of great value was the Johnsville human centrifuge. With this device, we learned to control the spacecraft during the accelerations imposed by launch and reentry and learned muscle control to aid blood circulation and respiration in the acceleration environment. The acceleration buildup during the flight was considerably smoother than that experienced on the centrifuge and probably for this reason and for obvious psychological reasons, the g-forces were much easier to withstand during the flight than during the training missions.

One other phenomenon that was experienced on the centrifuge proved to be of great value during the flight. Quite often, as the centrifuge changed rapidly from a g-level, a false tumbling sensation was encountered. This became a common and expected sensation and when the same thing occurred at launch vehicle cutoff, it was in no way disturbing. A quick glance at my instruments convinced me that I, indeed, was not tumbling.

The pilot's confidence comes from all the foregoing training methods and many other areas, but the real confidence comes from participation in the day-to-day engineering decisions and testing that occur during the pre-flight check-out at Cape Canaveral. It was during this time that I learned the particular idiosyncrasies of the spacecraft I was to fly. A great deal of time had already been spent in learning both normal and emergency system operations. But during the testing at the pre-flight complex and at the launching pad, I learned all the differences between this spacecraft and the simulator that had been used for training. I learned the various noises and vibrations that are connected with the operation of the systems. This was the time that I really began to feel at home in this cockpit. This training was very beneficial on launch day because I felt that I knew this spacecraft and what it would do, and having spent so much time in the cockpit I felt it was normal to be there.

As a group, we astronauts feel that after the spacecraft arrives at the Cape, our time is best spent in participating in spacecraft activities. This causes some conflict in training, since predicting the time test runs of the pre-flight checkouts will start or end is a mystic art that is understood by few and is unreliable at its best. Quite frequently this causes training sessions to be cancelled or delayed, but it should be of no great concern since most of the training has been accomplished prior to this time. The use of the trainers during this period is primarily to keep performance at a peak and the time required will vary from pilot to pilot.

At the time the spacecraft is moved from the pre-flight complex to the launching pad, practically all training stops. From this time on, I was at the pad full time participating in or observing every test that was made on the spacecraft – launch-vehicle combination. Here, I became familiar with the launch procedure and grew to know and respect the launch crew. I gained confidence in their professional approach to and execution of the pre-launch tests.

THE FLIGHT

On the day of the flight I followed the following schedule:

Event	a.m. e.s.t.
Awakened	1.10
Breakfast	1:25
Physical examination	1:55
Sensors attached	2:25
Suited up	2:35
Suit pressure check	3:05
Entered transfer van	3:30
Arrived at pad	3:55
Manned the spacecraft	3:58
Launched	7:20

As can be seen, 6 hours and 10 minutes elapsed from the time I was awakened until launch. This time is approximately evenly divided between activities prior to my reaching the pad and time I spent at the pad. In this case, we were planning on a launch at 6:00 a.m., e.s.t., but it will probably always be normal to expect some holds that cannot be predicted. While this time element appears to be excessive, we can find no way to reduce it below this minimum at the present. Efforts are still continuing to reduce the pre-countdown time so that the pilot will not have had an almost full working day prior to liftoff.

After insertion in the spacecraft, the launch countdown proceeded smoothly and on schedule until T-45 minutes when a hold was called to install a misaligned bolt in the egress hatch.

After a hold of 30 minutes, the countdown was resumed and proceeded to T-30 minutes when a brief hold was called to turn off the pad searchlights. By this time, it was daylight; and the lights, which cause interference with launch-vehicle telemetry, were no longer needed.

One more hold was called at T-15 minutes to await better cloud conditions because the long focal length cameras would not have been able to obtain proper coverage through the existing overcast.

After holding for 41 minutes, the count was resumed and proceeded smoothly to liftoff at 7:20 a.m., e.s.t.

The communications and flow of information prior to liftoff were very good. After participating in the pre-launch test and the cancellation two days previously, I was very familiar with the countdown and knew exactly what was going on at all times.

As the Blockhouse Capsule Communicator (Cap Com) called ignition, I felt the launch vehicle start to vibrate and could hear the engines start. Just seconds after this the time-elapsed clock started and the Mercury Control Center Cap Com confirmed liftoff. At that time I punched the Time Zero Override, started the stopwatch function on the spacecraft clock, and reported that the elapsed-time clock had started.

The powered flight portion of the mission was in general very smooth. A low-order vibration started at approximately T+50 seconds, but it did not develop above a low level and was undetectable after about T+70 seconds. This vibration was in no way disturbing and it did not cause interference in either communications or vision. The magnitude of the accelerations corresponds well to the launch simulations on the centrifuge, but the onset was much smoother.

Communications throughout the powered flight were satisfactory. The VOX (voice operated relay) was used for pilot transmissions instead of the push-to-talk button. The noise level was never high enough at any time to key the transmitter. Each standard report was made on time and there was never any requirement for myself or the Cap Com to repeat any transmission.

Vision out the window was good at all times during launch. As viewed from the pad, the sky was its normal light blue; but as the altitude increased, the sky became a darker and darker blue until approximately two minutes after liftoff, which corresponds to an altitude of approximately 100,000 feet, the sky rapidly changed to an absolute black. At this time, I saw what appeared to be one rather faint star in the center of the window. It was about equal in brightness to Polaris. Later, it was determined that this was the planet Venus whose brightness is equal to a star of magnitude of minus three.

Launch-vehicle engine cutoff was sudden and I could not sense any tail-off of the launch vehicle. I did feel, as I described earlier, a very brief tumbling sensation. The firing of the escape-tower clamp ring and escape rocket is quite audible and I could see the escape rocket motor and tower throughout its tail-off burning phase and for what seemed like quite some time after that climbing off to my right. Actually, I think I was still watching the tower at the time the posigrade rockets fired, which occurred ten seconds after cutoff. The tower was still definable as a long, slender object against the black sky at this time.

The posigrade firing is a very audible bang and a definite kick, producing a deceleration of approximately 1g. Prior to this time, the spacecraft was quite stable, with no apparent motion. As the posigrade rockets separated the spacecraft from the launch vehicle, the spacecraft angular motions and angular accelerations were quite apparent. Spacecraft damping which was to begin immediately after separation was apparently satisfactory, although I cannot really report on the magnitude of any angular rates caused by posigrade firing.

The spacecraft turnaround to retro-fire attitude is quite a weird maneuver to ride through. At first, I thought the spacecraft might be tumbling out of control. A quick check of the instruments indicated that turnaround was proceeding much as those experienced on the procedures trainer, with the exception of roll attitude which appeared to be very slow and behind the schedule that I was expecting.

As the turnaround started, I could see a bright shaft of light, similar to the sun shining into a blackened room, start to move from my lower left up across my torso. Even though I knew the window reduces light transmissions equivalent to the Earth's atmosphere, I was concerned that it might shine directly into my eyes and blind me. The light moved across my torso and disappeared completely.

A quick look through the periscope after it extended did not provide me with any useful information. I was unable to see land, only clouds and the ocean.

The view through the window became quite spectacular as the horizon came into view. The sight was truly breathtaking. The Earth was very bright, the sky was black, and the curvature of the Earth was quite prominent. Beneath the Earth and sky, there was a border which started at the Earth as a light blue and became increasingly darker with altitude. There was a transition region between the dark blue and the black sky that is best described as a fuzzy grey area. This is a very narrow band, but there is no sharp transition from blue to black. The whole border appeared to be uniform in height over the approximately 1,000 miles of horizon that was visible to me.

The Earth itself was very bright. The only landmark I was able to identify during the first portion of the weightlessness period was the Gulf of Mexico coastline between Apalachicola, Florida, and Mobile, Alabama. The cloud coverage was quite extensive and the curvature of this portion of the coast was very difficult to distinguish. The water and land masses were both a hazy blue, with the land being somewhat darker. There was a frontal system south of this area that was clearly defined.

One other section of the Florida coast came into view during the left yaw maneuver, but it was a small section of beach with no identifiable landmarks.

The spacecraft automatic stabilization and control system (ASCS) had made the turn-around maneuver from the position on the launch vehicle to retro-fire attitude. The pitch and yaw axes stabilized with only a moderate amount of overshoot as predicted, but the roll attitude was still being programmed and was off by approximately 15° when I switched from the autopilot to the manual proportional control system. The switchover occurred ten minutes later than planned to give the ASCS more time to stabilize the space-craft. At this point, I realized I would have to hurry my programmed pitch, yaw and roll maneuvers. I tried to hurry the pitch-up maneuver; I controlled the roll attitude back within limits, but the view out the window had distracted me, resulting in an overshoot in pitch. This put me behind my schedule even more. I hit the planned yaw rate but overshot in yaw attitude again. I realized that my time for control maneuvers was up and I decided at this point to skip the planned roll maneuver, since the roll axis had been exercised during the two previous maneuvers, and go immediately to the next task.

This was the part of the flight to which I had been looking forward. There was a full minute that was programmed for observing the Earth. My observations during this period have already been reported in this paper, but the control task was quite easy when only the horizon was used as a reference. The task was somewhat complicated during this phase, as a result of lack of yaw reference. This lack was not a problem after retrofire when Cape Canaveral came into view. I do not believe yaw attitude will be a problem in orbital flight because there should be ample time to pick adequate checkpoints; even breaks in cloud formations would be sufficient.

The retro-sequence started automatically and at the time it started, I was slightly behind schedule. At this point, I was working quite hard to get into a good retro-fire attitude so that I could fire the retro-rockets manually. I received the countdown to fire from Mercury Control Center Cap Com and fired the retro-rockets manually. The retro-rockets, like the escape rockets and posigrades, could be heard quite clearly. The thrust buildup was rapid and smooth. As the first retro-rocket fired, I was looking out the window and I could see that a definite yaw to the right was starting. I had planned to control the spacecraft attitude during retro-fire by using the horizon as a reference, but as soon as the right yaw started, I switched my reference to the flight instruments. I had been using my instruments during my retro-fire practice for the two weeks prior to the launch in the Cape Canaveral procedures training since the activity at the Cape prevented the use of the ALFA trainer located at the NASA-Langley Air Force Base. This probably explains the instinctive switch to the flight instruments.

The retro-fire difficulty was about equal to the more severe cases that have been presented on the procedures trainer.

Immediately after retro-fire, Cape Canaveral came into view. It was quite easy to identify. The Banana and Indian Rivers were easy to distinguish and the white beach all along the coast was quite prominent. The colors that were the most prominent were the blue of the ocean, the brownish-green of the interior, and the white in between, which was obviously the beach and surf. I could see the building area on Cape Canaveral. I do not recall being able to distinguish individual buildings, but it was obvious that it was an area where buildings and structures had been erected.

Immediately after retro-fire, the retro-jettison switch was placed in the armed position, and the control mode was switched to the rate command control system. I made a rapid check to ascertain that the system was working in all axes and then I switched from the UHF transmitter to the HF transmitter.

This one attempt to communicate on HF was unsuccessful. At approximately peak altitude, the HF transmitter was turned on and the UHF transmitter was turned off. All three receivers – UHF, HF, and emergency voice – were on continuously. Immediately after I reported switching to HF, the Mercury Control Center started transmitting to me on HF only. I did not receive any transmission during this period. After allowing the HF transmitter approximately ten seconds to warm up, I transmitted but received no acknowledgement that I was being received. Actually, the Atlantic Ship telemetry vessel located in the landing area and the Grand Bahama Island did receive my HF transmissions. Prior to the flight, both stations had been instructed not to transmit on the assigned frequencies unless they were called by the pilot. After switching back to the UHF transmitter, I received a call on the emergency voice that was loud and clear. UHF transmissions were satisfactory throughout the flight. I was in continuous contact with some facility at all times, with the exception of a brief period on HF.

Even though all communications equipment operated properly, I felt that I was hurrying all transmissions too much. All of the sights, sounds, and events were of such importance that I felt compelled to talk of everything at once. It was a difficult choice to decide what was the most important to report at any one time. I wanted as much as possible recorded so that I would not have to rely on my memory so much for later reporting.

As previously mentioned, the control mode was switched from manual proportional to rate command immediately after retro-fire. The procedures trainer simulation in this system seems to be slightly more difficult than the actual case. I found attitudes were easy to maintain and rates were no problem. The rate command system was much easier to fly than the manual proportional system. The reverse is normally true on the trainer. The sluggish roll system was probably complicating the control task during the manual proportional control phase of the flight, while roll accelerations appeared to be normal on the rate command system.

The rate command control system was used after retro-fire and throughout the reentry phase of the flight. At the zero rate command position, the stick was centered. This system had a deadband of plus or minus 3 degrees per second. Our experience on the procedures trainer had indicated that this system was more difficult to fly than the manual proportional control system. This was not the case during this flight. Zero rates and flight attitudes were easy to maintain. The records do indicate that an excessive amount of fuel was expended during this period. Approximately 15 percent of the manual fuel supply was used during the two minutes the system was operating. A major portion of the two-minute period was during the reentry when thrusters were operating almost continuously to dampen the reentry oscillations.

The 0.5 g telelight illuminated on schedule and shortly thereafter I reported g's starting to build. I checked the accelerometer and the g level was something less than 1 g at this time. The next time I reported, I was at 6 g and I continued to report and function throughout the high-g portion of the flight.

The spacecraft rates increased during the reentry, indicating that the spacecraft was oscillating in both yaw and pitch. I made a few control inputs at this time, but I could not see any effects on the rates, so I decided just to ride out the oscillations. The pitch rate needle was oscillating full scale at a rapid rate of plus or minus 6 degrees per second during this time and the yaw rate began oscillating full scale slightly later than pitch. At no time were these oscillations noticeable inside the spacecraft.

During this phase of reentry, and until main parachute deployment, there is a noticeable roar and a mild buffeting of the spacecraft. This is probably the noise of a blunt object moving rapidly through the atmosphere and the buffeting is not distracting nor does it interfere with pilot function.

The drogue parachute deployment is quite visible from inside the spacecraft and the firing of the drogue chute mortar is clearly audible. The opening shock of the drogue parachute is mild; there is a mild pulsation or breathing of the drogue parachute which can be felt inside the spacecraft.

As the drogue parachute is released, the spacecraft starts to drop at a greater rate. The change in g-field is quite noticeable. Main parachute deployment is visible out the window also. A mild shock is felt as the main parachute deploys in its reefed condition. The complete parachute is visible at this time. As the reefing cutters fire, the parachute deploys to its fully opened condition. Again, a mild shock is felt. About 80 percent of the parachute is visible at this time and it is quite a comforting sight. The spacecraft rotates and swings slowly under the parachute at first; the rates are mild and hardly noticeable.

The spacecraft landing in the water was a mild jolt; not hard enough to cause discomfort or disorientation. The spacecraft recovery section went under the water and I had the feeling that I was on my left side and slightly head down. The window was completely covered with water and there was a disconcerting gurgling noise. A quick check showed no water entering the spacecraft. The spacecraft started to slowly right itself; as soon as I was sure the recovery section was out of the water, I ejected the reserve parachute by actuating the recovery aids switch. The spacecraft then righted itself rapidly.

I felt that I was in good condition at this point and started to prepare myself for egress. I had previously opened the face plate and had disconnected the visor seal hose while descending on the main parachute. The next moves in order were to disconnect the oxygen outlet hose at the helmet, release the chest strap, release the lap belt and shoulder harness, release the knee straps, disconnect the biomedical sensors, and roll up the neck dam. The neck dam is a rubber diaphragm that is fastened on the exterior of the suit, below the helmet attaching ring. After the helmet is disconnected, the neck dam is rolled around the ring and up around the neck, similar to a turtle-neck sweater. This left me connected to the spacecraft at two points, the oxygen inlet hose which I needed for cooling and the helmet communications lead.

At this time, I turned my attention to the door. First, I released the restraining wires at both ends and tossed them towards my feet. Then I removed the knife from the door and placed it in the survival pack. The next task was to remove the cover and safety pin from the hatch detonator. I felt at this time that everything had gone nearly perfectly and that I would go ahead and mark the switch position as had been requested.

After about three or four minutes, I instructed the helicopter to come on in and hook onto the spacecraft and confirmed the egress procedures with him. I unhooked my oxygen inlet hose and was lying on the couch, waiting for the helicopter's call to blow the hatch. I was lying flat on my back at this time and I had turned my attention to the knife in the survival pack, wondering if there might be some way I could carry it out with me as a souvenir. I heard the hatch blow – the noise was a dull thud – and looked up to see blue sky out the hatch and water start to spill over the doorsill. Just a few minutes before, I had gone over egress procedures in my mind and I reacted instinctively. I lifted the helmet from my head and dropped it, reached for the right side of the instrument panel, and pulled myself through the hatch.

After I was in the water and away from the spacecraft I noticed a line from the dye marker can over my shoulder. The spacecraft was obviously sinking and I was concerned that I might be pulled down with it. I freed myself from the line and noticed that I was floating with my shoulders above water.

The helicopter was on top of the spacecraft at this time with all three of its landing gear in the water. I thought the copilot was having difficulty hooking onto the spacecraft and I swam the four or five feet to give him some help. Actually, he had cut the antenna and hooked the spacecraft in record time.

The helicopter pulled up and away from me with the spacecraft and I saw the personal sling start down: then the sling was pulled back into the helicopter and it started to move away from me. At this time, I knew that a second helicopter had been assigned to pick me up, so I started to swim away from the primary helicopter. I apparently got caught in the rotor wash between the two helicopters because I could not get close to the second helicopter, even though I could see the copilot in the door with a horse collar swinging in the water. I finally reached the horse collar and by this time, I was getting quite exhausted. When I first got into the water, I was floating quite high up; I would say my armpits were just about at the water level. But the neck dam was not up tight and I had forgotten to lock the oxygen inlet port; so the air was gradually seeping out of my suit. Probably the most air was going out around the neck dam, but I could see that I was gradually sinking lower and lower in the water and was having a difficult time staying afloat. Before the copilot finally got the horse collar to me, I was going under water quite often. The mild swells we were having were breaking over my head and I was swallowing some salt water. As I reached the horse collar, I slipped into it and I knew that I had it on backwards; but I gave the 'up' signal and held on because I knew that I wasn't likely to slip out of the ring. As soon as I got into the helicopter, my first thought was to get on a life preserver so that if anything happened to the helicopter, I wouldn't have another ordeal in the water. Shortly after this time, the copilot informed me that the spacecraft had been dropped as a result of an engine malfunction in the primary helicopter.

POST-FLIGHT

The post-flight medical examination onboard the carrier was brief and without incident. The loss of the spacecraft was a great blow to me, but I felt that I had completed the flight and recovery with no ill effects.

The post-flight medical debriefing at the Grand Bahama Island installation was thorough and complete. The demands on me were not unreasonable.

CONCLUSIONS

From the pilot's point of view the conclusions reached from the second U.S. manned suborbital flight are as follows:

(1) The manual proportional control system functioned adequately on this flight. The system is capable of controlling the retro-fire accurately and safely. The roll axis is underpowered and causes some difficulty. The rate command system functioned very well during this flight. All rates were damped satisfactorily, and it is easy to hold and maintain the attitudes with the rate command system. If the rate of fuel consumption that was experienced on this flight is true in all cases, it would not be advisable to use the rate control system during ordinary orbital flights to control attitudes. It should be used only for retro-fire and reentry. The autopilot functioned properly with the possible exception of the five seconds of damping immediately after separation. This period is so brief that it was impossible to determine the extent of any damping. The turnaround maneuver in the pitch and yaw axes was approximately as predicted, but the roll axis was slow to respond.

(2) The pilot's best friend on the orbital flight is going to be the window. Out this window, I feel he will be able to ascertain accurately his position at all times. I am sure he will be able to see stars on the dark side and possibly on the daylight side, with a little time to adapt the eyes. The brighter stars and planets will certainly be visible.

(3) Spacecraft rates and oscillations are very easy to ascertain by looking at the horizon and ground check points. I feel that drift rates will be easy to distinguish on an orbital flight when there is time to concentrate on specific points outside the window.

(4) Sounds of pyrotechnics, control nozzles, and control solenoids are one of the pilot's best cues as to what is going on in the spacecraft and in the sequencing. The sounds of posigrades, retro-rockets, and mortar firing are so prominent that these become the primary cues that the event has occurred. The spacecraft telelight panel becomes of secondary importance and merely confirms that a sequence has happened on time. The sequence panel's main value is telling the pilot when an event should have occurred and has not.

(5) Vibrations throughout the flight were of a low order and were not disturbing. The buffeting at maximum dynamic pressure and a Mach number of 1 on launch was mild and did not interfere with pilot functions. Communications and vision were satisfactory throughout this period. The mild buffeting on reentry does not interfere with any pilot functions.

(6) Communications throughout the flight were satisfactory. Contact was maintained with some facility at all times. There was never any requirement to repeat a transmission.

(7) During the flight, all spacecraft systems appeared to function properly. There was no requirement to override any system. Every event occurred on time and as planned.

Appendix 2: Voice communications to/from Mercury-Redstone 4 (MR-4) *Liberty Bell 7*

All times in mission elapsed time (MET)

00:00:01 CAPCOM [Al Shepard]:	Liftoff.
00:00:03 Grissom:	*Ah, roger. This is Liberty Bell 7. The clock is operating.*
00:00:08 CAPCOM:	Loud and clear, José, don't cry too much.
00:00:11 Grissom:	*Oke-doke.*
00:00:18 Grissom:	*Okay, it's a nice ride up to now.*
00:00:20 CAPCOM:	Loud and clear.
00:00:21 Grissom:	*Roger.*
00:00:28 Grissom:	*Okay; the fuel is go; about one and a half g's; cabin pressure is just coming off the peg; the O2 is go; we have 26 amps.*
00:00:36 CAPCOM:	Roger. Pitch 88 [degrees], the trajectory is good.
00:00:39 Grissom:	*Roger, looks good here.*
00:00:54 Grissom:	*OK there, we're starting to pick up a little noise and vibration; not bad, though, at all.*
00:01:01 Grissom:	*OK, the fuel is go; one and a half g's; cabin is eight [psi] ; the O2 is go; 27 amps.*
00:01:08 Grissom:	*And... [transmission lost]*
00:01:09 CAPCOM:	Pitch is...[transmission lost]
00:01:10 Grissom:	*Four g's, five g's...[transmission lost]*
00:01:11 CAPCOM:	Pitch is 77 [degrees]; trajectory is go.
00:01:13 Grissom:	*Roger. Cabin pressure is about six [psi] and dropping slightly. Looks like she's going to hold about 5.5 [psi].*
00:01:23 Grissom:	*Eh... [transmission lost].*
00:01:24 Grissom:	*Cabin... [transmission lost].*
00:01:24 Grissom:	*Believe me, O2 is go.*
00:01:26 CAPCOM:	Cabin pressure holding 5.5 [psi].
00:01:29 Grissom:	*Roger, roger.*

(continued)

C. Burgess, *Liberty Bell 7: The Suborbital Mercury Flight of Virgil I. Grissom*, Springer Praxis Books, DOI 10.1007/978-3-319-04391-3, © Springer International Publishing Switzerland 2014

(continued)

00:01:31 Grissom:	*This is Liberty Bell 7. Fuel is go; two and a half g's; cabin pressure is 5.5 [psi]; O2 is go; main [bus] 25 [volts], isolated - ah, isolated [bus] is 28 [volts]. We are go.*
00:01:46 CAPCOM:	Roger. Pitch is 62 [degrees]; trajectory is go.
00:01:49 Grissom:	*Roger. It looks good in here.*
00:01:56 Grissom:	*Everything is good; cabin pressure is holding; suit pressure is okay; two minutes and we got four g's; fuel is go; ah, feel the hand controller move just a hair there; cabin pressure is holding; O2 is go; 25 amps.*
00:02:15 CAPCOM:	Roger, we have a go here.
00:02:16 Grissom:	*And I see a star!*
00:02:17 CAPCOM:	Stand by for cutoff.
00:02:23 Grissom:	*There went the tower.*
00:02:24 Chase 1:	Roger, there went the tower, affirmative Chase.
00:02:26 Grissom:	*Roger, squibs are off.*
00:02:31 CAPCOM:	Roger.
00:02:33 Grissom:	*There went the posigrades, capsule has separated. We are at zero-g and turning around and the sun is really bright.*
00:02:42 CAPCOM:	Roger cap. sep. [capsule separation light] is green; turnaround has started, manual handle out.
00:02:47 Grissom:	*Oh, boy! Manual handle is out; the sky is very black; the capsule is coming around into orbit attitude; the roll is a little bit slow.*
00:03:01 CAPCOM:	Roger.
00:03:02 Grissom:	*I haven't seen the booster anyplace. Okay, rate command is coming on. I'm in orbit attitude, I'm pitching up. Okay, 40... [transmission lost] ... Wait, I've lost some roll here someplace.*
00:03:10 CAPCOM:	Roger, rate is command coming on. You're trying manual pitch.
00:03:15 Grissom:	*OK, I got roll back. OK, I'm at 24 [degrees] in pitch.*
00:03:20 CAPCOM:	Roger, your IP [impact point] is right on, Gus, right on.
00:03:24 Grissom:	*Okay. I'm having a little trouble with rate, ah, with the manual control.*
00:03:28 CAPCOM:	Roger.
00:03:31 Grissom:	*If I can get her stabilized here, all axes are working all right.*
00:03:36 CAPCOM:	Roger. Understand manual control is good.
00:03:40 Grissom:	*Roger, it's -- it's sort of sluggish, more than I expected.*
00:03:45 Grissom:	*Okay, I'm yawing.*
00:03:47 CAPCOM:	Roger, yaw.
00:03:50 Grissom:	*Left, ah.*
00:03:51 Grissom:	*Okay, coming back in yaw. I'm a little bit late there.*
00:03:57 CAPCOM:	Roger. Reading you loud and clear, Gus.
00:03:59 Grissom:	*Lot of stuff -- there's a lot of stuff floating around up here.*
00:04:02 Grissom:	*I'm going to skip the yaw [maneuver], ah, or [rather the] roll [maneuver] because I'm a bit late and I'm going to try this rough yaw maneuver. About all I can really see is clouds. I haven't seen any land yet.*
00:04:15 CAPCOM:	Roger, you're on the window. Are you trying a yaw maneuver?

(continued)

(continued)

00:04:18 Grissom:	*I'm trying the yaw maneuver and I'm on the window. It's such a fascinating view out the window and you can't help but look out that way.*
00:04:25 CAPCOM:	I understand.
00:04:29 Grissom:	*You, sh, ah, really. There I see the coast, I see.*
00:04:30 CAPCOM:	Four plus thirty [elapsed time since launch], Gus.
00:04:37 Grissom:	*I can see the coast, but I can't identify anything.*
00:04:42 CAPCOM:	Roger, four plus thirty [elapsed time since launch], Gus.
00:04:44 Grissom:	*Okay, let me get back here to retro attitude, retro sequence has started.*
00:04:48 CAPCOM:	Roger, retro sequence has started. Go to retro attitude.
00:04:52 Grissom:	*Right, we'll see if I'm in bad, not in very good shape here.*
00:04:57 CAPCOM:	Got 15 seconds, plenty of time, I'll give you a mark at 5:10 [elapsed time since launch].
00:05:01 Grissom:	*OK, retro attitude [light] is still green.*
00:05:05 CAPCOM:	Retros on my mark, three, two, one, mark.
00:05:11 Grissom:	*OK, there's one firing, there's one firing.*
00:05:13 CAPCOM:	Roger, retro one.
00:05:19 Grissom:	*There's two firing, nice little boost. There went three.*
00:05:21 CAPCOM:	Roger, three, all retros are fired.
00:05:23 Grissom:	*Roger, roger.*
00:05:25 Grissom:	*OK, yeah, they're fired out right there.*
00:05:29 CAPCOM:	Roger, retro jettison armed.
00:05:33 Grissom:	*Retro jettison is armed, retro jettison is armed, going to rate command.*
00:05:36 Grissom:	*OK, I'm going to switch.*
00:05:39 CAPCOM:	Roger. Understand manual fuel handle is in.
00:05:41 Grissom:	*Manual fuel handle is in, mark, going to HF.*
00:05:44 CAPCOM:	Roger, HF.
00:05:52 CAPCOM:	Liberty Bell 7, this is CAPCOM on HF, 1, 2, 3, 4, 5. How do you read, Seven?
00:06:00 Grissom:	*I got you.*
00:06:05 CAPCOM:	This is CAPCOM on HF, 1, 2, 3, 4, 5. How do you read, Seven?
00:06:08 Grissom:	*...here, do you read me, do you read me on HF? ...going back to U[HF]... Boy, is that... retro, I'm back on UHF and, ah, the jett-- the retros have jettisoned. Now I can see the Cape and, oh boy, that's some sight. I can't see too much.*
00:06:34 Grissom:	*Roger, I am on UHF high, do you read me?*
00:06:38 CAPCOM:	Roger, reading you loud and clear UHF high, can you confirm retro jettison?
00:06:41 Grissom:	*OK, periscope is retracting, going to reentry attitude.*
00:06:47 CAPCOM:	Roger, retros have jettisoned, scope has retracted, you're going to reentry attitude.
00:06:51 Grissom:	*Affirmative.*
00:06:56 CAPCOM:	Bell 7 from CAPCOM, your IP [impact point] is right on.
00:07:00 Grissom:	*Roger, I'm in reentry attitude.*
00:07:05 Grissom:	*Ah.*

(continued)

(continued)

00:07:07 CAPCOM:	Roger, how does it look out the window now?
00:07:09 Grissom:	*Ah, the sun is coming in and so all I can really see is just, ah, just darkness, the sky is very black.*
00:07:14 CAPCOM:	Roger, you have some more time to look if you like.
00:07:27 CAPCOM:	Seven from CAPCOM, how do you feel up there?
00:07:30 Grissom:	*I feel very good, auto fuel is 90 [percent], manual fuel is 50 [percent].*
00:07:33 CAPCOM:	Roger, 0.05 g in ten [seconds].
00:07:37 Grissom:	*OK.*
00:07:50 Grissom:	*OK, everything is very good, ah.*
00:07:54 Grissom:	*I have 0.05 g [light] and roll rate has started.*
00:07:57 CAPCOM:	Roger.
00:08:03 Grissom:	*Got a pitch rate in here, OK, g's are starting to build.*
00:08:09 CAPCOM:	Reading you loud and clear.
00:08:11 Grissom:	*Roger, g's are building, we're up to six [g's]*
00:08:17 Grissom:	*There's nine [g's]*
00:08:19 Grissom:	*There's about ten [g's]; the handle is out from under it; here I got a little pitch rate coming back down through seven [g's].*
00:08:32 CAPCOM:	Roger, still sound good.
00:08:34 Grissom:	*Okay, the altimeter is active at 65 [thousand feet]. There's 60 [thousand feet].*
00:08:38 CAPCOM:	Roger, 65,000 [feet].
00:08:42 Grissom:	*OK, I'm getting some contrails, evidently shock wave, 50,000 feet; I'm feeling good. I'm very good, everything is fine.*
00:08:49 CAPCOM:	Roger, 50,000 [feet].
00:08:52 Grissom:	*45,000 [feet], do you still read?*
00:08:54 CAPCOM:	Affirmative, still reading you. You sound good.
00:09:00 Grissom:	*Okay, 45,000 feet, do you read?*
00:09:07 Grissom:	*35,000 feet, if you read me.*
00:09:19 Grissom:	*30,000 feet, everything is good, everything is good.*
00:09:24 CAPCOM:	Bell 7, this is CAPCOM. How… [transmission lost]
00:09:28 Grissom:	*25,000 feet.*
00:09:36 Grissom:	*Approaching drogue chute altitude.*
00:09:41 Grissom:	*There's the drogue chute. The periscope has extended.*
00:09:45 CAPCOM:	This is…we have a green drogue [light] here, Seven, how do you read?
00:09:49 Grissom:	*OK, we're down to 15,000 feet, if anyone reads. We're on emergency flow rate, can see out the periscope okay. The drogue chute is good.*
00:10:03 CAPCOM:	Roger, understand drogue is good, the periscope is out.
00:10:05 Grissom:	*There's 13,000 feet.*
00:10:09 CAPCOM:	Roger.
00:10:14 Grissom:	*There goes the main chute; it's reefed; main chute is good; main chute is good; rate of descent is coming down, coming down to -- there's 40 feet per second, 30 feet per, 32 feet per second on the main chute, and the landing bag is out green.*

(continued)

(continued)

00:10:40 Grissom:	*Hello, does anyone read Liberty Bell? Main chute is good, landing bag [light] is on green.*
XX:XX:XX CAPCOM:	…and the landing bag [light] is on green.
XX:XX:XX Atlantic Ship:	Liberty Bell 7, Liberty Bell 7, this is Atlantic Ship CAPCOM. Read you loud and clear, our telemetry confirms your events, over.
XX:XX:XX Grissom:	*Ah, roger, is anyone reading Liberty Bell 7? Over.*
XX:XX:XX Cardfile 23:	Roger, Liberty Bell 7, reading you loud and clear. This is Cardfile 23. Over.
00:10:52 Grissom:	*Atlantic Ship CAPCOM, this is Liberty Bell 7, how do you read me? Over.*
XX:XX:XX Atlantic Ship:	Read you loud and clear, loud and clear. Over. Liberty Bell 7, Liberty Bell 7, this is Atlantic Ship CAPCOM. How do you read me? Over.
00:11:12 Grissom:	*Atlantic Ship CAPCOM, this is Liberty Bell 7, I read you loud and clear, how me? Over.*
XX:XX:XX Atlantic Ship:	Roger, Bell 7, read you loud and clear, your status looks good, your systems look good, we confirm your events. Over.
00:11:28 Grissom:	*Ah, roger, and confirm the fuel has dumped. Over*
00:11:34 Atlantic Ship:	Roger, confirm again, confirm again, has your auto fuel dumped?
00:11:42 Grissom:	*Auto fuel and manual fuel has dumped.*
XX:XX:XX Atlantic Ship:	Roger, roger.
00:11:47 Grissom:	*And I'm in the process of putting the pins back in the door at this time.*
00:12:04 Grissom:	*OK, I'm passing through 6,000 feet, everything is good, ah.*
00:12:15 Grissom:	*I'm going to open my Faceplate.*
00:12:35 Grissom:	*Hello, I can't get one back in; I can't get one door pin back in. I've tried and tried and can't get it back in. And I'm coming ATS [Atlantic Ship], I'm passing through 5,000 feet and I don't think I have one of the door pins in.*
XX:XX:XX Atlantic Ship:	Roger, Bell 7, roger.
00:13:04 Grissom:	*Do you have any word from the recovery troops?*
XX:XX:XX Cardfile 23:	Liberty Bell 7, this is Cardfile 23; we are heading directly toward you.
00:13:18 Grissom:	*ATS, this is CAP -- this is Libery Bell 7. Do you have any word from the recovery troops?*
XX:XX:XX Atlantic Ship:	Negative, Bell 7, negative. Do you have any transmission to MCC [Mercury Control Center]? Over.
00:13:33 Grissom:	*Ah, roger, you might make a note that there is one small hole in my chute. It looks like it is about six inches by six inches -- it's sort of a -- actually it's a triangular rip, I guess.*
XX:XX:XX Atlantic Ship:	Roger, roger.

(continued)

(continued)

00:13:49 Grissom:	*I'm passing through 3,000 feet; and all the fuses are in flight conditions; ASCS is normal, auto; we're on rate command; gyros are normal; auto retro jettison is armed; squibs are armed also. Four fuel handles are in; decompress and recompress are in; retro delay is normal; retro heat is off; cabin lights are both. TM [telemeter] is on. Rescue aids is auto; landing bag is auto; retract scope is in auto; retro attitude is in auto. All the three, five pull rings are in. Going down through some clouds to 2,000 feet. ATS, I'm at 2000 feet; everything is normal.*
XX:XX:XX Atlantic Ship:	Roger, Bell 7, what is your rate of descent again? Over.
00:14:39 Grissom:	*The rate of descent is varying between 28 and 30 feet per second.*
XX:XX:XX Atlantic Ship:	Ah roger, roger, and once again, verify your fuel has dumped, over.
00:14:54 Grissom:	*OK, my max g was about 10.2; altimeter is 1,000 [feet]; cabin pressure is approaching 15 [psi]. Temperature is 90 [degrees Fahrenheit]. Coolant quantity is 30 [percent]; temperature is 68 [degrees Fahrenheit]; pressure is 14 [psi]; main 02 is 60 [percent]; normal is, main is 60 [percent]; emergency is 100 [percent]; suit fan is normal; cabin fan is normal. We have 21 amps, and I'm getting ready for impact here. Can see the water coming right on up.*
XX:XX:XX Atlantic Ship:	Liberty Bell 7, Liberty Bell 7, this is Atlantic CAPCOM, do you read me? Over.
XX:XX:XX Grissom:	*OK, does anyone read Liberty Bell 7? Over.*
XX:XX:XX Hunt Club 1:	Liberty Bell 7, Hunt Club 1 is two miles southwest of you.
XX:XX:XX Cardfile 9:	Liberty Bell 7, this is 9 Cardfile. We have your entry into the water. Will be over you in just about 30 seconds.
00:16:35 Grissom:	*Roger, my condition is good. OK, the capsule is floating, slowly coming vertical, have actuated the rescue aids. The reserve chute has jettisoned, in fact I can see it in the water, and the whip antenna should be up.*
XX:XX:XX Grissom:	*Hunt Club, do you copy?*
XX:XX:XX Grissom:	*OK, Hunt Club, this is... Don't forget the antenna.*
XX:XX:XX Hunt Club 1:	This is Hunt Club, say again.
00:18:07 Grissom:	*Hunt Club, this is Liberty Bell 7. My antenna should be up.*
XX:XX:XX Hunt Club 1:	This is Hunt Club 1...your antenna is erected.
00:18:16 Grissom:	*Ah, roger.*
00:18:23 Grissom:	*OK, give me how much longer it'll be before you get here.*
XX:XX:XX Hunt Club 1:	This is Hunt Club 1, we are in orbit now at this time, around the capsule.
00:18:32 Grissom:	*Roger, give me about another five minutes here to mark these switch positions here, before I give you a call to come in and hook on. Are you ready to come in and hook on anytime?*
XX:XX:XX Hunt Club 1:	Hunt Club 1, roger we are ready any time you are.
00:18:44 Grissom:	*OK, give me another three or four minutes here to take these switch positions, then I'll be ready for you.*
XX:XX:XX Hunt Club 1:	One, wilco.

(continued)

(continued)

XX:XX:XX Card File 9:	Hey Hunt Clubs, Card File, Card File 9, I'll stand by to escort you back as soon as you lift out. I'll keep other aircraft at at least 2,000 feet.
XX:XX:XX Hunt Club 1:	Ah, Bell 7, this is Hunt Club 1.
00:20:15 Grissom:	*Go ahead, Hunt Club 1.*
XX:XX:XX Hunt Club 1:	Roger, this is One, observe something, possibly the canister in the water alongside the capsule. Will we be interfering with any TM [telemetry] if we come down and take a look at it?
00:20:26 Grissom:	*Negative, not at all, I'm just going to put the rest of this stuff on tape and then I'll be ready for you, in just about two more minutes, I would say.*
XX:XX:XX Hunt Club 1:	One, roger.
XX:XX:XX CAPCOM:	Liberty Bell 7, CAPCOM at the Cape on a test count, over.
XX:XX:XX CAPCOM:	Liberty Bell 7, Cape CAPCOM on a test count, over.
XX:XX:XX Card File 9:	Any Hunt Club, this is 9 Card File.
XX:XX:XX Hunt Club 1:	Station calling Hunt Club, say again.
00:24:03 Card File 9:	This is Niner Card File, there's an object on a line in the water, ah, just about 160 degrees. The NASA people suspect it's the dye marker that didn't activate; ah, say it's about, ah, three-fourth of a mile out from the capsule. Ah, after the lift out, will you take a check on it? Over.
XX:XX:XX Hunt Club 1:	Ah, this is Hunt Club 1, roger. Will have Hunt Club 3 check at this time. You copy, Three?
XX:XX:XX Hunt Club 3:	Hunt Club 1, believe he said three-fourth of a mile?
XX:XX:XX Card File 9:	This is 9 Card, that is affirmative.
00:25:20 Grissom:	*OK, Hunt Club. This is Liberty Bell 7. Are you ready for the pickup?*
00:25:26 Hunt Club 1:	This is Hunt Club 1. This is affirmative.
00:25:30 Grissom:	*OK, latch on, then give me a call and I'll power down and blow the hatch, OK?*
00:25:36 Hunt Club 1:	This is Hunt Club 1, roger. Will give you a call when we're ready to blow.
00:25:42 Grissom:	*Roger, I've unplugged my suit so I'm kinda warm now so.*
00:25:46 Hunt Club 1:	One, roger.
00:25:52 Grissom:	*Now if you tell me to, ah, you're ready for me to blow, I'll have to take my helmet off, power down, then blow the hatch.*
00:25:59 Hunt Club 1:	One, roger, and when you blow the hatch, the collar will already be down there waiting for you, and we're turning base at this time.
00:26:09 Grissom:	*Ah, roger.*

At this time Grissom powered down *Liberty Bell 7*, ending communications from the spacecraft.

Appendix 3: Flight plan for Mercury-Redstone (MR-4) Liberty Bell 7

Time:min:sec	Event
0:00	Liftoff
0:30	Systems report
1:00	Systems report
1:15	Cabin pressure report
1:30	Systems report
2:00	Systems report
2:23	Launch-vehicle engine cutoff
	Tower jettison
	Retro-jettison switch to OFF
2:33	Spacecraft separation from launch vehicle
2:38	Spacecraft turnaround to flight attitude on autopilot
3:00	Transfer of flight control from autopilot to manual proportional control system, and evaluation of system
4:00	Spacecraft yawed to 45 degrees to left using horizon as attitude reference
5:10	Retrograde rockets fired manually
5:35	Retro-jettison system armed
	Transfer of flight control from manual proportional control system to rate command control system
6:10	Radio transmitter switched from UHF to HF
6:40	Retro-package jettison
	Periscope retracts automatically
7:00	Spacecraft positioned into reentry attitude
7:46	Communications switched back to UHF transmitter
9:41	Reentry starts
	Drogue parachute deploys
	Snorkels open
10:13	Emergency rate oxygen flow
15:37	Main parachute deployment
	Landing

C. Burgess, *Liberty Bell 7: The Suborbital Mercury Flight of Virgil I. Grissom*, Springer Praxis Books, DOI 10.1007/978-3-319-04391-3, © Springer International Publishing Switzerland 2014

VIRGIL I. ('GUS') GRISSOM'S HONORS AND AWARDS

- Posthumously awarded the Congressional Space Medal of Honor, 1978
- Posthumously made Honorary Mayor of the City of Newport News, Virginia
- Distinguished Flying Cross for service in Korea
- Air Medal with cluster for service in Korea
- Army Good Conduct Medal
- American Campaign Medal
- World War Two Victory Medal
- National Defense Service Medal with star
- Korean Service Medal
- United Nations Service Medal
- Two NASA Distinguished Service Medals
- The NASA Exceptional Service Medal
- The Air Force Command Astronaut Wings
- Honorary Doctorate, Florida Institute of Technology

TRIBUTES TO GUS GRISSOM

- Grissom Air Reserve Base in Indiana is named in his honor
- The Virgil I. Grissom Library in the Denbigh section of Newport News, Virginia is named after him
- *CSI* character Gil Grissom was named after him
- *Thunderbirds* character Virgil Tracy was named after him
- Grissom Hall at Purdue University is named after him
- Grissom Hall at SUNY Fredonia is named after him
- Grissom Hall at Florida Institute of Technology is named after him
- Gus Grissom is Class Exemplar of the U.S. Air Force Academy's class of 2007
- Named on the Space Memorial Mirror at the KSC Visitor Center, Florida
- Grissom Crater on the Moon
- Grissom Hill on the planet Mars
- 2161 Grissom is a main belt asteroid
- The airport in Bedford, Indiana was renamed Virgil I. Grissom Municipal Airport
- Virgil "Gus" Grissom Park in Fullerton, California
- The Gus Grissom Stakes thoroughbred horse race is run annually at Hoosier Park, Anderson, Indiana
- Grissom Island located off Long Beach, California

In September 2011, Gus Grissom's son Scott and brother Lowell attended a base rededication ceremony at Grissom Air Reserve Base, Indiana. (Photo: U.S. Air Force/Tech. Sgt. Mark R.W. Orders-Woempner)

SCHOOLS NAMED AFTER GUS GRISSOM

- Virgil I. Grissom High School, Huntsville, Alabama
- Virgil I. Grissom Middle School, Mishawaka, Indiana
- Grissom Elementary School, Gary, Indiana
- Grissom Elementary School, Mincie, Indiana
- Virgil I. Grissom Middle School, Tinley Park, Illinois
- Virgil I. Grissom Elementary School, Hegewisch community of Chicago, Illinois
- Virgil I. Grissom Middle School, Sterling Heights, Michigan
- Virgil I. Grissom School No. 7, Rochester, New York
- Virgil I. Grissom Junior High School 226, Queens, New York
- Virgil Grissom Elementary School, Princeton, Iowa
- Grissom Elementary School, Tulsa, Oklahoma
- Virgil I. Grissom Elementary School, Houston, Texas
- Virgil Grissom Elementary School, Old Bridge, New Jersey
- V.I. Grissom Elementary School in the now-closed Clark Air Base, Philippines

Editorial from the *New York Times* newspaper, 22 July 1961

> **Across the Frontiers of Space**
>
> Virgil I. Grissom of the Air Force rode a Redstone rocket high into the skies from Cape Canaveral yesterday to become the third human being to surge into space and to harness the powerful blast of the rocket to the purposes of man. The achievement, marred only by the loss of the capsule, is a worthy one to write beside the deeds of the pioneers who opened a continent, conquered the plains and breached the ramparts of the Rockies.
>
> To Grissom and his breed, and to the scientists, technicians and engineers who made his flight possible, the towering cumulus of the heavens, the twinkling radiance of a star, present the same challenge that has always stirred the hearts of explorers – the mystery of the unknown.
>
> By such deeds, by such skills, by such teamwork and by such faith, Man plumbs the future, rolls back the infinite and excites the imagination and aspiration of all whose reach would exceed their grasp.

C. Burgess, *Liberty Bell 7: The Suborbital Mercury Flight of Virgil I. Grissom*, Springer Praxis Books, 267
DOI 10.1007/978-3-319-04391-3, © Springer International Publishing Switzerland 2014

About the author

Australian author Colin Burgess grew up in Sydney's southern suburbs. Initially working in the wages department of a major Sydney afternoon newspaper (where he first picked up his writing bug) and as a sales representative for a precious metals company, he subsequently joined Qantas Airways as a passenger handling agent in 1970 and two years later transferred to the airline's cabin crew. He would retire from Qantas as an onboard Customer Service Manager in 2002, after 32 years' service. During those flying years several of his books on the Australian prisoner-of-war experience and the first of his biographical books on space explorers such as Australian payload specialist Dr. Paul Scully-Power and teacher-in-space Christa McAuliffe had already been published. He has also written extensively on spaceflight subjects for astronomy and space-related magazines in Australia, the United Kingdom and the Unites States.

In 2003 the University of Nebraska Press appointed him series editor for the ongoing *Outward Odyssey* series of 12 books detailing the entire social history of space exploration, and was involved in co-writing three of these volumes. His first Springer-Praxis book, *NASA's Scientist-Astronauts*, co-authored with British-based space historian David J. Shayler, was released in 2007. *Liberty Bell 7: The Suborbital Mercury Flight of Virgil I. Grissom* will be his seventh title with Springer-Praxis, and a follow-on companion to his earlier work, *Freedom 7: The Historic Flight of Alan B. Shepard, Jr.* He is currently researching further space-related subjects for future publication. He regularly attends astronaut functions in the United States and is well known to many of the pioneering space explorers, allowing him to conduct personal interviews for these books.

Colin and his wife Patricia still live just south of Sydney. They have two grown sons, two grandsons and a granddaughter.

C. Burgess, *Liberty Bell 7: The Suborbital Mercury Flight of Virgil I. Grissom*, Springer Praxis Books,   269
DOI 10.1007/978-3-319-04391-3, © Springer International Publishing Switzerland 2014

Index